China's Role in the Burma War:
Jet Fuel, Money, and Weapons

BY ANTONIO GRACEFFO, PHD, MBA, M.A. (ECONOMICS)

Webantonio67@gmail.com

China's Role in the Burma War: Jet Fuel, Money, and Weapons

First published November 2025

By Mary Labita Press

Copyright © 2025 Antonio Graceffo

ISBN: 979-8-9897950-3-1

TABLE OF CONTENTS

DEDICATION

This book is dedicated to the people of Burma. May God bring peace to their country, and may the people defeat the junta and achieve the federal democracy they long for—free from Beijing's influence.

ACKNOWLEDGEMENTS

Special thanks to David Eubank and the Free Burma Rangers (FBR), who were instrumental in providing resources and access in Burma. Deep gratitude also goes to the Catholic Church, the resistance forces, and the Burmese diaspora in the churches of Chiang Mai, who shared valuable information and encouraged me to keep writing.

AUTHOR'S NOTE

This short book, which examines China's extensive involvement in the ongoing Burma civil war, is based on the master's thesis that I presented on December 1, 2024 for my master's degree in International Relations and Global Security at American Military University (AMU). The thesis title was *Assessing China's Role in the Burma Conflict: Jet Fuel, Money, and Weapons Perpetuating an Endless War.* This version contains all of the original sources and citations, but some of the information has been updated.

My research confirms Beijing's financial, political, and military support for the Burmese junta and select Ethnic Armed Organizations (EAOs). It demonstrates that China's interventions align with its strategic objectives, including advancing the Belt and Road Initiative and strengthening its regional influence. Utilizing a qualitative methodology, I analyzed primary sources such as government documents, official statements, and reports from aid organizations, supplemented by secondary literature from established experts. Data on Chinese investments, forced displacements, and the humanitarian crisis illustrate the detrimental impact of China's support on Burma's population, while evidence of Chinese military aid and sightings of PLA troops further validate China's active involvement. My findings confirm that Beijing's support has significantly prolonged the conflict, countering the intended effects of international sanctions on the junta. The research concludes that China's actions are primarily driven by long-term security and economic interests, presenting substantial challenges to US and Indian national security in the Indo-Pacific region.

INTRODUCTION

The war in Burma, also known as Myanmar, has persisted for over 70 years, making it the world's longest-running conflict. Since gaining independence in 1948, the central government has faced ongoing battles with various ethnic armed organizations (EAOs) formed by many of Myanmar's 135 ethnic groups (Bertrand et al., 2022). These EAOs have frequently shifted alliances and engaged in periodic ceasefires (Ong, 2023). However, since the 2021 military coup, fighting has intensified, with EAOs increasingly joining forces to challenge the junta, and some targeting Chinese investments due to Beijing's ties with the military (Chan, 2024). China plays a complex role in the conflict, supporting the junta to safeguard its strategic interests while maintaining relations with certain EAOs to protect its investments (Ong, 2023).

China's influence on both state and non-state actors in Burma is a crucial factor in understanding the broader implications for regional and global security (Ganesan, 2011; Asia News Monitor, 2023). Burma is a key component of China's global strategy, particularly through the China-Myanmar Economic Corridor, which includes vital oil and gas pipelines and rail links from Yunnan province to Burma's Indian Ocean ports (Kobayashi and King, 2022). Securing this energy infrastructure, and solving the "Malacca Dilemma" — China's vulnerability due to most of its crude oil imports passing through the Strait of Malacca chokepoint — is essential for Beijing before it risks a potential conflict with the United States, such as an invasion of Taiwan (Ong, 2023; Myers, 2023).

This book examines China's complex involvement in the ongoing Burma war, focusing on its financial support of the ruling junta, its backing of certain EAOs, and its role in the Rohingya genocide. Beijing's actions raise important questions about China's broader strategy in the context of globalization, regional security, and the evolving roles of state and non-state actors. Assessing China as a self-interested actor, more concerned with securing its economic and strategic interests than with the humanitarian or political stability of Burma, this analysis seeks to illustrate how China's interventions reflect these strategic goals, while shedding light on its regional influence and the implications for both Burma and the international community.

ARMED GROUPS

A more detailed and comprehensive annotated list of armed groups can be found in Appendix A at the end of this book.

THE ARMED LANDSCAPE: MYANMAR'S COMPLEX WEB OF RESISTANCE

In 1948, the Karen National Union (KNU) emerged as the first ethnic armed organization (EAO) fighting against the Burmese government, marking the beginning of what would become the world's longest-running civil war. With Myanmar comprising 135 distinct ethnic groups, this initial conflict has evolved into a complex web of armed resistance that spans nearly eight decades.

The February 1, 2021 military coup fundamentally transformed this landscape. What had been a collection of mostly localized ethnic conflicts suddenly became a nationwide uprising against military rule. The coup shattered years of fragile peace negotiations and prompted many groups that had signed the government's Nationwide Ceasefire Agreement to resume armed resistance.

THE SCALE OF ARMED OPPOSITION

As of 2025, over 120 armed organizations actively oppose the military junta, representing the largest coordinated resistance in the country's modern history. These range from massive ethnic armies with tens of thousands of fighters to local village defense units with a few dozen members. The resistance operates across three main categories:

1) Major Ethnic Armed Organizations: Long-established groups like the Arakan Army (45,000+ troops), Kachin Independence Army (20,000), and Karen National Liberation Army (15,000) that have fought for decades seeking autonomy or independence for their ethnic regions.

2) People's Defense Forces (PDFs): The largest category, with an estimated 100,000+ fighters nationwide. These are the armed wing of the exiled National Unity Government (NUG), formed by lawmakers elected in 2020 who were ousted in the coup. PDFs operate in 250 of Myanmar's 330 townships.

3) Post-Coup Formations: New groups like the Karenni Nationalities Defense Force (7,000+ troops) and Chinland Defense Force that emerged specifically to resist the 2021 coup.

THE THREE BROTHERHOOD ALLIANCE

The most significant development since the coup has been the formation of powerful military alliances. The Three Brotherhood Alliance — comprising the Arakan Army, Myanmar National Democratic Alliance Army (MNDAA), and Ta'ang National Liberation Army (TNLA) — launched Operation 1027 in October 2023, capturing dozens of military bases and fundamentally shifting the balance of power.

TERRITORIAL CONTROL

The junta now controls an estimated 30% to 40% of Myanmar's territory, primarily urban centers and transport corridors. Opposition forces control 35% to 45%, with the remainder contested or under autonomous arrangements. This represents a dramatic reversal from pre-coup conditions when the military dominated most of the country.

THE NEUTRALITY QUESTION

Not all armed groups oppose the junta. The powerful United Wa State Army (UWSA, 30,000 troops) maintains official neutrality while allegedly supplying weapons to resistance groups. Some smaller ethnic armies and border guard forces remain aligned with the military, creating a complex patchwork of shifting alliances.

INTERNATIONAL DIMENSIONS

The conflict increasingly involves regional powers. China maintains complex relationships with multiple ethnic armies along its border, sometimes pressuring for ceasefires to protect its economic interests. Meanwhile, weapons and support flow across porous borders with Thailand, India, and Bangladesh.

BACKGROUNDER ON THE HISTORY OF THE BURMA CONFLICT

The February 1, 2021 coup in Myanmar that toppled the democratically elected government of Aung San Suu Kyi made international headlines. However, as global attention waned, many remained unaware that this conflict is the world's oldest ongoing civil war, lasting over 75 years since Myanmar's independence from Britain in 1948. About 68% of the country's population are Burmans (Bamar), who have historically controlled the government and military; the latter is known as the Tatmadaw. The remaining 32% are divided among ethnic minorities who have endured decades of military oppression.

The road to Myanmar's current desperate state has been marked by successive military coups. General Ne Win seized power in 1962, establishing a one-party socialist state. In 1988, widespread protests led to another coup by General Saw Maung, who established the State Law and Order Restoration Council (SLORC), later rebranded as the State Peace and Development Council (SPDC) under General Than Shwe until 2011.

As Saw Taw Nee, head of Foreign Affairs for the KNU, explained from a military base in Karen State: "The military dictatorship started in 1962. I was born in 1964. Like 90% of Burmese, I was born into a war that has lasted longer than my lifetime."

THE CEASEFIRE PERIOD (2011-2021): A FRAGILE PEACE

From 2011 to 2021, Myanmar transitioned to a quasi-civilian government under the military-backed Union Solidarity and Development Party (USDP), followed by Aung San Suu Kyi's National League for Democracy (NLD) after their 2015 landslide victory. However, the military retained significant control under the 2008 constitution, including 25% of parliamentary seats and key ministries.

Before the 2021 coup, Myanmar's civil war had entered a period of partial stabilization through complex ceasefire arrangements. Between

1989 and 2008, 39 of the 40 bilateral ceasefires were fully observed. Under President Thein Sein's government (2011-2016), all of the 18 new ceasefires initiated during his term remained intact until the 2021 coup.

By 2021, Myanmar's ceasefire architecture had evolved into two main tracks. The oldest elements were **bilateral ceasefires**, agreements dating from 1989-1994. Many of these deals were purely military truces which aimed to freeze conflicts rather than resolve root causes. These included powerful groups like the UWSA, which governed parts of Shan State as a de facto autonomous entity. The **Nationwide Ceasefire Agreement** (NCA), signed on October 15, 2015 by eight EAOs — and joined by two more in 2018 — was signed by the KNU, the Restoration Council of Shan State (RCSS), Chin National Front (CNF), and others. The NCA was designed as a political ceasefire providing a path to dialogue on autonomy, federalism, and constitutional reform.

Despite these arrangements, only about 20% of EAOs ever joined the NCA. The remaining 80% — including the most powerful factions like the Kachin Independence Army, Arakan Army, and TNLA — viewed the agreement as an unacceptable demand for surrender. These groups instead formed the Federal Political Negotiation and Consultative Committee (FPNCC) in 2017 under UWSA leadership, positioning themselves as an ethnic-led peace coalition with Chinese backing.

This period created a fragile but enduring status quo where much of Myanmar's periphery was under effective EAO control, and the central government had accepted this reality in exchange for reduced hostilities.

THE COUP THAT CHANGED EVERYTHING (2021)

The February 1, 2021 military coup shattered this delicate balance. General Min Aung Hlaing seized power after the NLD's re-election victory in 2020, arresting Suu Kyi and declaring a nationwide state of emergency.

Before the coup, many in the Bamar ethnic majority viewed EAOs with deep suspicion. Decades of Tatmadaw propaganda had painted these groups as terrorists, drug lords, or separatists. The Bamar public had placed faith in elections and peaceful protest, seeing federalism as foreign concept associated only with distant border wars. The coup obliterated this perception. On February 2, 2021, the Civil Disobedience Movement (CDM) was born. What began with healthcare workers refusing to work under the junta exploded into a nationwide general strike. At its height, over 417,000 civil servants — more than 40% of the national bureaucracy — had joined the CDM, along with businesses, consumers, and citizens from every sector.

The military's response was ruthless. Protesters were gunned down in the streets, nighttime raids abducted dissenters, and reports of torture, mass detentions, and executions proliferated. This reign of terror radicalized the public. According to junta sources, 367 officials appointed by the State Administration Council (the name the junta gave itself until mid-2025) were assassinated between February 2021 and February 2022.

THE GREAT AWAKENING: BAMAR-ETHNIC UNITY

In a historic shift, thousands of Bamar activists fled cities and entered territories controlled by groups like the KNU, Karenni Nationalities Defense Force (KNDF), CNF, and Kachin Independence Army. New recruits from urban centers like Yangon, Bago, and Mandalay began undergoing military training in jungle camps, receiving instruction in both combat tactics and political education about federalism and ethnic histories.

Personal transformations unfolded across the country. Social media became spaces for Bamar individuals to issue public apologies to ethnic minorities. "In the past, I neglected to know about the [other] ethnic groups, their suffering and losses, and I acted like it wasn't my business," said Ma Wai, a young Bamar woman who joined the Burma People's Liberation Army. "I also didn't notice that I was privileged as a Bamar."

The dry zone of central Myanmar, the heartland of the Bamar Buddhist population and rarely touched by warfare, erupted in what some now call the "Anya War." The central plains became the site of insurgencies led by Bamar People's Defense Forces (PDFs) working with ethnic forces.

David Eubank, director of Free Burma Rangers and a former US special forces officer turned missionary, has worked in Myanmar for nearly 30 years. He noted that EAOs were not fooled by the Tatmadaw's offer to stay out of the coup. When the coup happened and reinforcements were sent to ethnic areas, Karen leaders declared: "The ceasefire is over."

THE CURRENT WAR: A NEW PHASE

Since the coup, reported clashes have surged by 67%. The conflict is characterized by ongoing, low-intensity warfare that occasionally escalates into intense fighting. However, the October 2023 launch of Operation 1027 by the Three Brotherhood Alliance appears to signal a fundamental shift, with resistance forces making substantial territorial gains and achieving numerous victories.

What has fundamentally changed since the coup is unprecedented cooperation. For the first time in modern history, ethnic groups and Bamar are working together under the National Unity Government (NUG), formed by ousted lawmakers and including representatives from ethnic minorities. The NUG's armed wing, the PDF, is working with ethnic armies, creating the largest coordinated resistance in Myanmar's history.

The rebels now control an estimated 60% to 70% of Myanmar's territory, though the Tatmadaw still holds major cities. Despite near economic collapse, the military continues fighting, bolstered by Chinese investment and Russian weapons. The war's outcome may ultimately depend on whether resistance forces can counter the junta's air superiority, which is their primary remaining advantage.

Myanmar's Human Catastrophe

As of October 2024, an estimated 3.5 million people were internally displaced within Myanmar, a staggering increase from pre-coup levels. By 2025, some 19.9 million people were in need of humanitarian assistance. The crisis now extends far beyond Myanmar's borders, with approximately 1 million Rohingya refugees remaining in Bangladesh in camps in Cox's Bazar and on Bhasan Char Island.

Thailand hosts an estimated 3-4 million Burmese migrants and refugees, though this figure represents a very rough estimate given that many migrants lack legal documentation and official identity papers. The 2,416-kilometer border between the two countries is porous, and sees constant movement back and forth, making precise counts nearly impossible. Of this population, approximately 90,000 refugees from Myanmar live in nine official refugee camps along the Thai-Myanmar border, primarily housing ethnic minorities such as the Karen and Karenni who have sought refuge there since the 1980s. The vast majority of the remaining Burmese population works throughout Thailand in a variety of legal circumstances, from documented workers under bilateral agreements to undocumented migrants vulnerable to arrest and deportation.

In 2024 alone, nearly 900,000 people were forced to flee their homes in Myanmar—a 37% increase from 2023. Human displacement and systematic bombing of civilian targets have left vast agricultural lands uncultivated, creating acute food insecurity even in major cities. An estimated 120,000 men, women, and children have been subjected to trafficking into "scam compounds" operating in border areas, where they face forced criminality, sexual exploitation, and organ harvesting.

The March 28, 2025 earthquake that struck central Myanmar further compounded the crisis, killing over 3,300 people and adding thousands more to the displaced population, overwhelming an already decimated healthcare system.

These consequences show that the coup did more than destroy Myanmar's fragile democratic experiment. It also destroyed the myth that ethnic resistance was someone else's war. The brutal repression exposed the military's violence as systemic, not situational. For the first time in decades, young Bamar men and women stand shoulder to shoulder with the same ethnic fighters their government once called terrorists. They're no longer protesting for change, but fighting for it.

LITERATURE REVIEW

Findley and Marineau (2015) confirm that China and Chinese entities have played both sides in the conflict, with Chinese citizens and companies purchasing forest products and resources directly from the rebels. Similarly, Ahram and King (2012) discuss China's involvement with Khun Sa (1934-2007), also known by his Chinese name, Zhang Qifu, one of the most powerful warlords in Burma, as an example of China's intervention in the conflict by supporting both EAOs and government forces to further Chinese interests.

Operating in Northern Shan State, bordering China's Yunnan province, Khun Sa led the Shan United Army, which later became the Mong Tai Army (MTA), fighting against the Burmese junta. Khun Sa was also known as the king of Burma's heroin trade and the largest heroin trafficker in the world, maintaining direct contacts with Chinese transnational crime organizations. To ensure trade routes and stability in Burma, Beijing began supplying the Burmese army with advanced weaponry to combat Khun Sa's forces. Eventually, ethnic Wa militias, who were already receiving support from China, and Karen militias, who had entered into a ceasefire agreement with the government, joined forces with the Burmese army and overran Khun Sa's stronghold (Ahram and King, 2012).

After Khun Sa surrendered to the Myanmar government in 1996, the MTA was disbanded. Many of its remnants subsequently reorganized into the Shan State Army-North (SSA-N) and Shan State Army-South (SSA-S). The latter has maintained a ceasefire with the government but continues to trade with China and collect taxes on government trade with China.

Earlier in his career, Khun Sa had been imprisoned for several years but was eventually released in an exchange after Shan forces kidnapped two Soviet specialists in 1973 (Ahram and King, 2012; Mandelkorn,

2020). This event marked the end of Soviet influence and ushered in the era of Beijing's dominance in Myanmar's civil war.

Zou and Fan (2019) explore the concept of Burma as a client state of China, much like Cuba was a client state of the Soviet Union during the Cold War. Shortly after gaining independence in 1948, Burma became the first non-communist country to recognize the People's Republic of China (PRC). Under the leadership of Prime Minister U Nu, Burma adopted a non-aligned foreign policy during the Cold War, seeking to balance its relations between the West and the Communist Bloc. U Nu was cautious of China's influence but understood the necessity of maintaining peaceful diplomatic ties with its powerful neighbor. At the same time, he maintained good relations with the West, accepting economic and military aid, particularly from the United States, while avoiding full alignment with either side. U Nu's neutral stance aimed to preserve Burma's independence and sovereignty, navigating the complex geopolitics of the era without fully committing to the interests of China or the West.

At the time, Burma was building a democratic socialist state that required significant financial aid to complete its socio-economic development plan, known as Pyidawtha. Yangon felt that the United States and the United Kingdom were not providing enough aid, although their contributions were substantial, and so it began to rely on China for support. US-Burma relations were further strained by the presence of Kuomintang (KMT) troops in northern Burma, who opposed the PRC, as well as by Yangon's growing alignment with Beijing. This created a cycle of distrust and violence, ultimately strengthening Sino-Burmese ties.

During the Korean War, Burma experienced an economic boom by selling exports to support the war. However, once the war ended, prices declined to pre-war levels, and Burma experienced a recession (Zou and Fan, 2019). This bust, combined with a split within the party and the Trade Union Council, further destabilized the political landscape. The leader of the Trade Union Council, Thakin Lwin, who also led the leftist wing of Burmese political power, followed a Communist Party line and adopted a pro-Soviet stance. Thakin Lwin objected to receiving aid from the British and denounced Burmese political leaders U Kyaw Nyein, who held a moderate policy toward both the West and the Soviet Union, and U Ba Swe, who sought to maintain a sound relationship with European countries.

Thakin Lwin pressured the Burmese government to adopt anti-imperialist and anti-colonialist attitudes against Western countries. British and American intelligence reports from the era suggested that Burmese Communists might collaborate with the Karen rebellion to overthrow U Nu's government; if not, Minister of Defense General Ne Win, was expected to lead a coup and usurp the civilian government. This demonstrates Beijing's long history of playing both sides, supporting the junta and certain EAOs, driven by China's interest in securing its own strategic goals rather than establishing democracy in Burma or improving the lives of the Burmese people.

The Cold War saw fierce competition in Burma between the Soviet Union, the PRC, and the United States. Mandelkorn (2020) viewed the civil war as a proxy war between the US and China, the first two Cold War powers to reach out to the newly independent Burma. When the socialist government of Burma effectively banned aid from the two major competitors, the Soviet Union took the initiative and began building infrastructure in Yangon and other parts of the country. At one point, the USSR had more influence and a greater presence in the capital and urban areas than either the US or the PRC. However, in the jungle provinces controlled by ethnic armed organizations, particularly those bordering China, the picture was quite different. The PRC and Thailand were the primary foreign influences, with PRC influence being more overt and directly tied to trade and investment.

Blue (2020) writes that China has three primary interests in Burma. First is security and uninterrupted cross-border trade with Yunnan province. China has always seen the country as crucial to its security, viewing Myanmar as strategically significant in terms of its peripheral areas. Diplomatic ties between the two nations were officially established in 1950. Over the past six decades, their relationship has been grounded in the Five Principles of Peaceful Coexistence, which were jointly agreed upon by Myanmar, China, and India in 1954 (Zhao, 2012). The PRC has traditionally felt most secure when Myanmar had a stable government, even if it was a military junta, as long as Western influence remained limited.

Next is economic cooperation, including increasing China's access to natural resources. Finally, China wants direct access to the Indian Ocean (Blue, 2020). The latter is necessary for China to secure its oil supplies from the Persian Gulf while bypassing the Malacca Strait. This led to China backing the mass killing and displacement of the Rohingya Muslim minority in Rakhine State, Burma. In 2011, under a quasi-civilian government, Burma defied Beijing by suspending the Chinese-backed Myitsone dam hydroelectric project. From the standpoint of the United States Marine Corps, Blue (2020) recognizes that China's engagement in Burma has significant implications for US national security in the Indo-Pacific, where China is deploying "multiple instruments of national power to pursue its strategic interests."

Energy is a top priority for China's resource interests. While Myanmar's hydrocarbon reserves may not be globally significant, they are crucial for China's energy security due to the region's gas reserves and Myanmar's strategic location. China has heavily invested in Myanmar's energy infrastructure, but before the 2021 coup, rising concerns about "new colonialism" and resource nationalism, along with competition from other global players like Japan and India, strained the relationship (Zhao, 2015). Under nominal democratic rule, Myanmar tried to reduce its dependence on China, but the COVID lockdowns, followed by the coup, drove out most other investors, leaving China as the regime's primary patron (Thein and Baron, 2024; Taunt, 2024).For China, Myanmar holds significant geopolitical importance due to its strategic location, providing access to the Indian Ocean and the Andaman Sea.

This is particularly crucial for China as it navigates instability in Pakistan, which threatens the China-Pakistan Economic Corridor — once considered the crown jewel of the Belt and Road Initiative, designed to address China's "Malacca Dilemma" by providing an alternative route for transporting oil from the Middle East (Zhao, 2015; Ahmad, 2023; Sahay, 2024; Myers, 2023). With Pakistan's instability and a Myanmar junta now more dependent on China and more compliant with Beijing's demands, the China-Myanmar Economic Corridor (CMEC) has grown in importance. Myanmar serves as a critical connection point in Southeast Asia, linking India and China, and playing a pivotal role in global trade routes such as the Asian Highway Networks and the Singapore-Kunming Rail Link (Zhao, 2015; Kyi, 2024).

While Myanmar's contribution to the region's gas supplies is modest compared to larger producers like Indonesia and Malaysia, its offshore fields — such as Yadana, Yetagun, and the upcoming Shwe fields — hold untapped potential. Chinese National Oil Companies (NOCs), including CNPC, Sinopec, and CNOOC, have been actively involved in oil and gas exploration in Myanmar. The China-Myanmar Oil and Gas Pipelines Project is particularly significant, featuring deep-water natural gas developments and onshore pipelines that transport oil and gas from western Myanmar to China (Zhao, 2015).

China's strategic Shwe gas and oil pipelines provide a direct energy route from the Indian Ocean to Yunnan in China. Running from Kyaukphyu in Myanmar to Ruili in China, the US$1.04 billion project is managed by the South-East Asia Pipeline Company (SEAP) and the

Shwe Consortium. Despite local protests, the pipeline bolsters China's energy security by bypassing the Strait of Malacca and ensuring a steady flow of natural gas from Myanmar's offshore fields (Zhao, 2015).

Parallel to the gas pipeline is a crude oil pipeline, also linking Myanmar to China's southwestern region. Built by China's CNPC, this pipeline transports oil from the Middle East and Africa to Myanmar's Maday Island, where new oil storage facilities and a deep-water port, funded by China, have been developed (Cameron, 2023). Myanmar provides security for the pipeline, while SEAP manages its operations (Zhao, 2015).

Beyond energy security, the pipelines are part of China's broader strategy to promote economic integration with Myanmar and Southeast Asia. The CMEC, developed along the pipeline routes, is expected to boost economic growth in Myanmar's provinces and enhance logistics between the two nations (Myers, 2020; Zhao, 2015).

For China, this project not only supports its underdeveloped southwestern provinces, such as Yunnan and Sichuan, but also extends its economic influence into Southeast Asia, addressing regional energy needs through improved infrastructure and refining capacity (Zhao, 2015).

Zhao (2012) explains that China has strategically filled the gaps left by Western sanctions in Myanmar, positioning its companies as key players in the gas sector and for future gas imports into China. As foreign direct investment in Myanmar has declined, China has emerged as a crucial investor and patron. Its two major sub-regional development initiatives, the Greater Mekong Sub-Region Development Scheme and Bangladesh-China-India-Myanmar regional economic cooperation, have been driving much of the economic activity between Yunnan and Myanmar. Previously, Thailand was the largest source of foreign direct investment in Myanmar, but China has now overtaken Bangkok. Additionally, Myanmar's trade deficit with China has been steadily growing (OEC, 2024). The Myanmar oil and gas projects will create a fourth route for China's energy imports, complementing ocean shipping via the Malacca Strait and pipelines from Kazakhstan and Russia. China has used diplomatic influence to foster relationships with oil and gas-exporting nations, including vetoing a UN Security Council resolution against Myanmar in 2007, after which Myanmar's junta awarded a major oil and gas contract to a PRC company despite being outbid by an Indian competitor (Zhao, 2012).

The Myitsone hydropower dam, a joint venture between China Power Investment Corporation (CPI) and Myanmar's military-linked Asia World, was poised to significantly boost China's influence in Myanmar. However, nationwide protests in 2011 forced the suspension of the project. It was expected to be the 15th largest dam in the world, generating 3,600 to 6,000 megawatts of electricity, with 90% going to China. The project became controversial due to concerns over environmental impact, displacement of local communities, and its historical significance. CPI's decision to negotiate with Myanmar's military government, bypassing the local Kachin organization, escalated tensions and contributed to the breakdown of a ceasefire between the Kachin and the government (Zhao, 2015).

In June 2024, Myanmar's junta, in need of funds, reopened negotiations with China to resume the Myitsone hydropower project. According to the new agreement, China will receive 80% of the electricity generated, the Myanmar government will receive 15%, and Asia World Company, which is building the dam, will receive 5% (Myitkyina Journal, 2024). Asia World Company has close ties to the UWSA, an ethnic armed group with strong connections to China (Rotberg, 1998).

The Letpadaung copper mine in Sagaing Region's Salingyi Township, a joint venture between China's Wanbao Mining (a subsidiary of NORINCO, a Chinese state-owned conglomerate with interests in arms manufacturing and mining) and Myanmar's Union of Myanmar Economic Holdings Ltd., has also faced protests since its planned expansion in 2011 (Amnesty International, 2017). Local villagers objected to the confiscation of over 7,800 acres of land and raised concerns about pollution in a nearby watershed. After a violent crackdown on demonstrators in 2012, a Commission of Inquiry led by Aung San Suu Kyi found that the project failed to meet international environmental standards or adequately compensate affected villagers. However, the report did not recommend halting the project but suggested improvements to meet international standards — a response that local activists deemed inadequate (Zhao, 2015).

Protests resulted in Letpadaung copper mine being shut down for several years. Militia members with ties to the government were hired to guard the mine. However, ethnic armed resistance forces detained 14 members of the military-backed Pyu Saw Htee militia as they traveled to the mine. The captives were handed over to the NUG, while a 20-year-old policeman guarding the mine surrendered to a separate resistance group, confirming that the junta had been training militiamen and police to operate heavy weaponry. Despite Wanbao's claim that operations had been suspended since the February 2021 coup, locals reported that activity at the site had resumed, and over 40 villagers were recently forced from their homes as the military occupied their community. Additionally, an NUG statement revealed that more than 400 homes in the area had been destroyed by junta arson attacks, and 17 civilians were killed by troops stationed at the mine (Business & Human Rights Resource Centre, 2024)

In 2024, the return of Chinese workers reignited protests, strikes, and military ambushes. Resistance fighters targeted the project, viewing it as a key source of income for the junta (RFA, 2024). After the attack, Tatmadaw units stationed nearby captured and tortured 23 local civilians (Myanmar Now, 2024). Both local people and EAOs blame China for enabling the junta through its investments. They also hold China responsible for the forced displacements, torture, and other suffering inflicted on local communities as the Tatmadaw clear the way for Chinese ventures.

Beijing maintains its strategic interests through its engagement with both the junta and certain EAOs. One example of an ethnic armed entity supporting China's investments is the Myanmar National Democratic Alliance Army (MNDAA), a member of a rebel alliance fighting against military rule. In 2024, the MNDAA pledged to protect Chinese nationals and investments in northern Myanmar, particularly as fighting over the town of Lashio in northern Shan State intensified (Asian News Monitor, 2024). Lashio is strategically located on the trade route connecting Myanmar with China, making it a vital hub for cross-border trade and economic activity. It is part of the CMEC, which links Myanmar's resources to China's southwestern provinces. Control of Lashio is crucial for maintaining stability in this economically significant area, where Chinese investments, such as oil and gas pipelines, are concentrated. The MNDAA's commitment to protect Chinese interests likely stems from the economic benefits that come with maintaining strong ties with China (CNI News, 2024).

Since the collapse of a China-negotiated ceasefire in July 2023, the MNDAA has claimed control of much of Lashio, though junta officials deny this. The MNDAA stated that it would consult allied organizations to prevent harm to Chinese investments and employees. Political analyst Than Soe Naing noted that rebel groups, including the TNLA and the Arakan Army, have previously committed to protecting Chinese interests under an earlier agreement, possibly due to pressure from Beijing (Asian News Monitor, 2024).

China's interests in northern Shan State include maintaining trade routes and border stability, as well as addressing online scam centers in the region. The FPNCC, an alliance of ethnic minority forces, has been entrusted by China to ensure these interests are safeguarded. In 2023, the committee called on China to help resolve the crisis triggered by Myanmar's military coup. Beijing has urged all parties in Myanmar to end hostilities and seek a peaceful resolution, although China's motivation for doing so derives from a need to protect its investments. Meanwhile, the UWSA, which is closely affiliated with China, has evacuated junta-affiliated prison staff trapped by fighting in Lashio, although the Wa army claims neutrality in the ongoing conflict (Asian News Monitor, 2024).

The FPNCC is made up of seven EAOs in Myanmar, including the UWSA, Arakan Army, Kachin Independence Army, and the MNDAA. Formed in 2017, the FPNCC serves as an alternative to the 2015 Nationwide Ceasefire Agreement (NCA), which was accepted by several EAOs but rejected by FPNCC members (BNI Online, 2024).

Internal conflicts within the FPNCC have intensified, particularly in northern Shan State, due to territorial disputes and inter-ethnic tensions. The Shan State Progress Party (SSPP) and the TNLA are at odds over control of Namkham Township, leading to military tensions after the TNLA expelled SSPP forces and replaced Shan-language signs with those in the Ta'ang language, sparking anger among Shan residents. Meanwhile, the TNLA's long-standing alliance with the Kachin Independence Army has also frayed, with the TNLA accusing the KIA of interfering in its territories and ordering the KIA to halt operations in certain areas. Similarly, tensions between the MNDAA and KIA have escalated after the MNDAA removed KIA flags and occupied Kachin-run educational institutions. Additionally, the MNDAA and TNLA, once joint partners in military operations, have clashed over territorial control in several regions. Ideologically, the UWSA has maintained neutrality in the conflict against the junta, while the KIA collaborates with the National Unity Government (NUG) and People's Defense Forces (PDF), further deepening internal divisions (BNI Online, 2024; Bynum, 2018). These internal conflicts highlight the diverse and often countervailing interests of EAOs. Beijing's efforts to secure a single, unified agreement with these groups to protect its investments are likely to face significant challenges, making such an outcome nearly impossible.

The FPNCC has strong ties to China, which often mediates in its peace talks. The lengthy list of internal conflicts between the various EAOs underscores the fact that a ceasefire is unlikely. It also demonstrates that while China wants stability and an end to fighting near its projects, it has no real interest in understanding or remedying the conflicts between the EAOs.

While the group is not formally aligned with the NUG, some FPNCC members have sheltered and trained anti-junta forces, such as the PDFs (DMG Newsroom, 2023). The FPNCC maintains relationships with China, primarily due to Beijing's interest in ensuring stability in northern Myanmar for trade and Belt and Road Initiative projects. Some members, like the UWSA, maintain ceasefires with the junta, while others, such as the MNDAA, frequently clash with military forces, revealing internal divisions and differing objectives within the alliance (BNI Online, 2024).

China's strategy in Myanmar has been marked by a delicate balance of engaging with all factions, particularly ethnic armed groups and the military junta, with the aim of maintaining border stability. China has played a key role in brokering ceasefires, such as in northern Shan State, to protect its economic interests, especially the CMEC. While China avoids taking sides, it has encouraged the UWSA and the FPNCC to remain engaged in peace efforts. The UWSA's recent involvement in Lashio highlights China's influence, as Beijing formally engages with the FPNCC as a negotiating body. However, China is cautious of the NUG, viewing it as aligned with Western democratic interests. Despite Beijing's ties to the junta and its backing of the Wa and Northern Alliance, China appears committed to a two-pronged approach — working with the junta to resolve the crisis through elections while maintaining strong ties with the EAOs to secure its broader regional goals. This strategy reflects China's desire to avoid instability while ensuring its long-term influence in Myanmar. (Das, 2024). This dynamic allows the FPNCC to negotiate with both the junta and China, while still supporting anti-junta forces, making it a complex player in Myanmar's conflict.

The most powerful EAO which supports China's interests is the UWSA, descendants of the original Burma Communist Party, which the PRC funded during the Cold War as part of a larger strategy to turn Burma into a communist ally. Beijing leverages its relations with the EAOs to further its interests when the junta is unresponsive or noncompliant. As China prioritizes border security, it works with both the Burmese junta and the UWSA to maintain control over the border. For this reason, Beijing has on several occasions attempted to negotiate a ceasefire.

Writing about why some EAOs enter into ceasefires while others do not, Dukalskis (2011) outlines how the PRC became the primary sponsor of the UWSA following the collapse of the Burma Communist Party. The SLORC (State Law and Order Restoration Council) reached a ceasefire agreement with the UWSA in 1989. This ceasefire granted the UWSA significant autonomy over the Wa region in northern Burma (Myanmar), allowing the group to retain its arms and operate with relative independence, particularly in areas along the China-Myanmar border. The agreement also permitted the UWSA to engage in cross-border trade, which became central to the economic activities of the Wa State.

The ceasefire was closely tied to China's interests, as Beijing had historically backed the Burma Communist Party, which had a strong presence in the Wa region. After withdrawing its support in the 1980s, China shifted its strategy toward stabilizing its border with Myanmar and maintaining influence in the region. The ceasefire between the SLORC and the UWSA, an ethnic Wa militia, aligned with China's broader goal of ensuring stability and protecting its economic and strategic interests along the border (Dukalskis, 2011).

The UWSA also benefited from economic ties with China, receiving supplies, trade opportunities, and even military equipment. These ties allowed China to maintain leverage over the region and indirectly influence developments in northern Myanmar without direct intervention (Dukalskis, 2011).

Fan and Zou (2023) raise an often forgotten wrinkle in the Sino-Burmese relationship: border disputes, which were not settled until the 1950s and 1960s. A clear delineation of the border was crucial to China at that time to contain the war and chaos in Myanmar, reduce transnational crime, and control cross-border trade. Additionally, Kuomintang (KMT) troops, who had settled in Burma after the Chinese Civil War, played a significant role in the armed ethnic conflict in Burma, and the PRC saw them as a security concern.

Jason Tower, country director for the Myanmar program at the US Institute of Peace, suggested that China appeared to be seeking a compromise where the military junta would cede some northern territories, while the rebel Three Brotherhood Alliance would scale back its war goals. He indicated that Beijing likely hoped the Tatmadaw would withdraw from certain areas, encouraging the alliance to shift its focus from eradicating the military dictatorship to stabilizing the region, particularly for Chinese projects. Tower explained that China was simultaneously helping the alliance achieve more limited political objectives while supporting the military by applying pressure on the alliance. He further noted that while China favored the alliance taking control of border regions, it was also pushing the group to engage with the military junta and limit its involvement in broader revolutionary activities. Additionally, Tower pointed out that China continues to collaborate with the junta on projects such as the Kyaukphyu Deep-

Sea Port in Rakhine State, which a more responsible and democratic Myanmar government might not have endorsed (Naing, 2024).

All of these sources confirm China's long-term, multifaceted involvement in the ongoing Burma war — through financial support of the junta, a state actor, as well as its backing of non-state EAOs. Additionally, these sources touched on China's culpability in the Rohingya genocide and explored how China's involvement in Myanmar fits within the context of globalization, regional security, and backing for Beijing's trade and resource interests. More research will be needed to quantify and more comprehensively answer the report question.

RESEARCH SINCE 2024-2025

Since mid-2024, China's involvement in the conflict has significantly deepened, though it remains carefully calibrated to protect Beijing's strategic interests. No credible evidence confirms the presence of Chinese military personnel inside Myanmar, yet China has substantially expanded its support of the junta through technology transfers, surveillance systems, private security companies, and strategic mediation efforts that prioritize economic stability over genuine peace.

CHINESE MILITARY SUPPORT AND TECHNOLOGY TRANSFERS

China's most visible contribution to the junta's military capabilities has been the massive expansion of drone warfare capabilities. Since mid-2024, Chinese-made UAV have become central to the junta's escalating air campaign against resistance forces. The military has established a dedicated Directorate of Drone Warfare under Brigadier-General Nay Myo Tun, acquiring thousands of Chinese UAVs adapted for combat and surveillance operations.

These acquisitions include loitering munitions, "suicide drones" (aka "kamikaze drones" modeled after Iran's Shahed-136, and advanced unmanned combat aerial vehicles documented in a January 2025 Janes report. Chinese companies such as Zhongyue Aviation Firefighting-Drone Company have reportedly entered into discussions with the junta to establish licensed production facilities inside Myanmar, representing a significant escalation in technology transfer.

China has also provided training for Burmese military personnel, with groups of junta air force pilots and drone operators receiving instruction in China. While no evidence confirms PRC trainers currently operating in-country, the scale and sophistication of the junta's drone warfare strongly suggests continued technical support, possibly including remote instruction or offshore training programs.

The state-owned Aviation Industry Corporation of China (AVIC) continues to supply Myanmar with fighter jets and transport aircraft, including the JF-17, FTC-2000G, and Y-8. Justice for Myanmar has confirmed that at least five AVIC aircraft types have been used in junta airstrikes. Troublingly, Airbus — partially owned by France, Spain, and Germany — still holds equity in AVIC and has even expanded partnerships, demonstrating how Western nations indirectly enable the junta's operations.

The Critical Role of Jet Fuel

The junta's devastating air campaign, responsible for over 4,000 airstrikes in the first three years post-coup, continues largely unimpeded due to steady jet fuel supplies despite international sanctions. Between March and April 2025 alone, the regime launched at least 140 air strikes, including in earthquake-hit areas of Sagaing and Mandalay, where a 7.7-magnitude earthquake on March 28, 2025 killed over 3,600 people.

While the US and Western allies have imposed jet fuel restrictions, enforcement remains critically weak. Mark Farmaner of Burma Campaign UK explains that "the Myanmar military doesn't have the ability to manufacture its own jet fuel" and that it continues to obtain supplies through Chinese and Vietnamese companies involved in indirect supply chains. Western governments have sanctioned only Burma-based entities while failing to target international suppliers.

The junta's fuel supply chain centers on the state-owned Myanmar Petrochemical Enterprise and crony conglomerates like Asia Sun Group. Amnesty International documented at least three aviation fuel shipments between January and June 2024, routed through Vietnam, Singapore, the UAE, and China via complex offshore transfers. PRC-owned tanker *Huitong 78* played a central role, while companies involved included Sahara Energy, CNOOC Singapore, and Royal Vopak.

At least five British insurance firms have underwritten sanctioned deliveries, facilitating the junta's fuel access. The UK could easily prevent its insurers from supporting sanctioned trades, as it did with Russian oil sanctions, but has failed to apply similar measures to Myanmar.

Chinese Private Security Companies (PSCs)

In February 2025, the junta passed the Private Security Services Law, legalizing foreign security company operations. The law permits PSCs provided their personnel are not active-duty foreign military members and 75% are Burmese citizens. This framework allows former PLA soldiers to operate under PSC banners, blurring lines between private contractors and Chinese state actors.

In October 2024, the junta formed a 13-member committee to establish a joint venture PSC with China, responsible for overseeing weapons imports and communications equipment. However, a memorandum of understanding remained unsigned as of April 2025. China Overseas Security Group, China's largest PSC, is likely involved in securing key infrastructure along China-Myanmar oil and gas pipelines.

While Chinese PSCs have raised concerns, their impact on active fighting remains minimal. Unlike Russia's Wagner Group or US-based Blackwater, Chinese PSCs are not structured for frontline combat. Their personnel generally lack combat experience and are not trained for offensive operations in Burma's predominantly jungle warfare environment. However, these companies significantly impact the junta's urban control capabilities, particularly digital surveillance and population monitoring. PSCs guard critical infrastructure that could naturally extend to data centers, surveillance hubs, and telecom nodes, effectively outsourcing repression while maintaining Chinese plausible deniability.

DIGITAL SURVEILLANCE AND REPRESSION

Burma's junta has constructed one of the world's most repressive digital control systems, powered by Chinese surveillance technology and biometric infrastructure. This represents a direct export of the PRC's digital authoritarianism model, transforming Burma into a testing ground for 21st-century state control.

China's famous "Great Firewall" is one of the world's most sophisticated internet censorship systems. A combination of facial recognition, AI-powered CCTV, and mobile phone surveillance enables real-time monitoring of citizens. Through its Digital Silk Road initiative, China has exported these technologies to authoritarian regimes, including Burma, where the military junta has adopted key elements of Chinese digital control infrastructure.

Following the coup, Burma's military rapidly expanded its control of the telecom sector, mandating surveillance system installation across all major networks. While interception systems were quietly introduced at Telenor as early as 2018, post-coup compliance became compulsory. These Chinese-modeled systems allow authorities to eavesdrop on calls, monitor private messages, and track user locations through cell tower triangulation.

High-level meetings between China's Ministry of Public Security and Burma's military officials have focused on strengthening security cooperation and facilitating surveillance technology rollout. Training sessions on biometric enrollment and management systems have been conducted across regions including Loikaw, Shan State, Magway, and Kayah.

The Unique Identification System

In 2023, the military launched a nationwide Unique Identification (UID) system based on a model similar to China's social credit system. While software origins remain unconfirmed, the junta has actively sought

PRC expertise, holding high-level meetings and visiting firms such as Beijing Hisign Technology, which specializes in facial and biometric recognition.

The UID system collects and links personal and biometric data, including names, addresses, phone numbers, fingerprints, iris scans, palm prints, and facial features, into a centralized database cross-referenced across platforms such as SIM card registration, financial services, and travel documentation. Registration is now mandatory for essential services including passport applications, banking, and border passes, giving the regime tighter control over emigration and population movement. During a September 2023 visit to Beijing by Burma's Minister for Immigration and Population, junta officials met with their Chinese counterparts to discuss integrating Chinese e-ID systems into Burma's immigration processes.

With China's support, the junta can now track movement, deny exit to dissidents, and arrest flagged individuals at passport offices and border checkpoints. The system makes it nearly impossible to flee using false identities, as biometric data is centrally stored and cross-referenced.

The junta's Person Scrutinization and Monitoring System (PSMS), reportedly developed with Huawei support, uses artificial intelligence and facial recognition for real-time individual monitoring. Since 2022, regime police have arrested over a thousand people using PSMS, with many released prisoners now placed on AI-powered watchlists.

Under "Safe City" initiatives, the junta in Naypyidaw has rolled out smart surveillance projects in the capital and other major cities including Yangon, Taunggyi, Bago, Mawlamyine, and Myitkyina. These areas operate under dense networks of AI-powered CCTV cameras, facial recognition software, and predictive policing algorithms that mirror China's own Safe City and Sharp Eyes programs.

CHINA'S GLOBAL SECURITY INITIATIVE AND BURMA

Burma is among several authoritarian governments that have signed onto Beijing's Global Security Initiative (GSI), a PRC-led framework promoting state-centric governance and control models. While framed as a commitment to peace and sovereignty, the GSI advances China's "cyber sovereignty" doctrine, encouraging total state control over digital infrastructure, communication, and data.

The Burmese junta revised the country's Cybersecurity Law to closely mirror China's own, embedding language about "sovereignty and stability" that reinforces a legal basis for sweeping surveillance and digital censorship. The junta's Ministry of Foreign Affairs has publicly endorsed the GSI, praising its emphasis on sovereignty and territorial integrity, code for non-interference in what Beijing and Naypyidaw define as internal affairs.

The junta has weaponized internet shutdowns to control information flow, prevent citizen organization, and block communication with resistance forces. At least 329 shutdowns have been imposed over four years, particularly in conflict zones like Chin, Sagaing, Rakhine, and Karenni. All 330 of the country's townships have experienced some form of communication blackout. According to Access Now and the #KeepItOn coalition, Burma led the world in intentional internet disruptions in 2024, with 85 documented shutdowns, 31 tied to human rights violations and 17 coinciding with military offensives. These disruptions, often timed with airstrikes, leave communities disconnected and unable to report attacks.

BEIJING'S CEASEFIRE MEDIATION STRATEGY

China has increasingly stepped in as mediator, brokering ceasefires between Myanmar's military and EAOs primarily to safeguard strategic interests in regional stability. However, these agreements repeatedly collapse, exposing underlying fragility and Beijing's transactional approach. Recent examples include negotiations with the TNLA in Yunnan and a January 2025 ceasefire with the MNDAA, the second such agreement in just over a year, as the January 2024 ceasefire collapsed within months.

Economic Motivations

China's primary motivation for brokering ceasefires is protecting economic and strategic interests — particularly resource extraction, infrastructure projects like the CMEC, and critical trade routes, rather than genuine peacekeeping. When clashes threaten key trade routes connecting Yunnan to Myanmar's ports, China intervenes to secure temporary truces.

However, once interests are safeguarded, Beijing exerts little pressure to ensure compliance. This explains why many China-brokered agreements fail: they do not resolve political grievances of EAOs but serve China's economic interests. The frequent collapse highlights fragility of China's mediation, driven by economic necessity rather than genuine commitment to resolving the civil war.

THE LATEST DEVELOPMENTS AS OF LATE NOVEMBER 2025

Recently, there have been numerous raids on scam centers that appear to be an attempt by both the Chinese Communist Party and the Burma junta to add a veneer of legitimacy to their economic and diplomatic ambitions by virtue signaling to the world that they support the international rules-based order and are taking a harsh stance on transnational crime. The junta's motivation is to deflect criticism of its flawed elections, which are scheduled to begin their first phase in December. For the Chinese Communist Party, the 2025 U.S.–China Commission report finds that the CCP is exploiting Burma (Myanmar) scam centers as a tool of espionage, using them to expand its diplomatic and security footprint and to obtain data on Americans.

According to the 2025 U.S.–China Economic and Security Review Commission report, because armed resistance groups are central to China's access to infrastructure and raw materials under the Belt and Road Initiative, Beijing tolerated the scam centers for years. But the scale of financial losses to Chinese citizens became a direct economic threat, prompting China to pressure the Burma military junta to shut them down.

Criminal syndicates have shifted from targeting Chinese citizens to Americans, allowing them to make more money while drawing less pressure from Beijing. In 2024, China reported a 30 percent drop in scam losses, while U.S. losses rose 40 percent. After China-led crackdowns, syndicates in Shwe Kokko began recruiting English-proficient workers to target Americans and Europeans.

Chinese criminals behind these scam centers have cultivated ties, some overt and others deniable, to the CCP. By supporting Belt and Road projects in Burma and spreading CCP propaganda overseas, scam bosses have turned their operations into tools of Beijing's diplomatic and economic expansion.

Chinese crime boss Wan Kuok-Koi, or "Broken Tooth," shows how Beijing tolerates criminal networks that serve its interests. After leading the 14K triad and serving 14 years in prison, he rebuilt his operations and, in 2019, partnered with the Karen Border Guard Force, an ethnic armed group aligned with the junta, to create the Dongmei Zone, marketed as tourism but quickly converted into a scam-center hub.

As he expanded his criminal empire, he rebranded himself as a pro-CCP figure and, in 2017, founded the World Hongmen History and Culture Association in Cambodia, which promotes CCP narratives on issues such as Hong Kong and Taiwan. He declared, "I used to fight for the cartels, and now I fight for the CCP," while laundering scam profits into businesses in China, including real estate and construction. Beijing has taken no action against him, signaling its willingness to tolerate criminal groups that support its broader agenda.

The CCP is also using the scam centers as a pretext to deploy its own security forces throughout the region. This inside-out strategy gives Beijing footholds inside the internal security apparatuses of neighboring countries. The Commission describes Southeast Asia as a pilot zone for tactics China intends to replicate in Africa, Latin America, and Central Asia.

Beijing has used the presence of Chinese transnational criminal groups to pressure governments to accept a larger Chinese security role. On May 25, 2023, China and Laos agreed to deepen law-enforcement and security cooperation to combat transnational crime. In September 2024, Cambodia's Minister of Interior Sar Sokha traveled to Beijing and committed to expanding joint law-enforcement efforts, particularly against criminal networks. In January 2025, China hosted a meeting of the Lancang–Mekong Law Enforcement Cooperation mechanism with Cambodia, Laos, Burma, Thailand, and Vietnam, where the parties agreed to increase intelligence sharing and joint operations targeting scam centers.

China has also used the scam-center issue to pressure Thailand, a U.S. treaty ally that historically resisted allowing Chinese police to operate on its soil, to permit Chinese security forces to work inside the country. After Chinese actor Wang Xing was abducted in Thailand and trafficked into a Burma-based scam center in January 2025, Chinese tourism to Thailand reportedly dropped by 33 percent, dealing a severe blow to Thailand's economy.

Seeking to reassure Chinese visitors, Thai Prime Minister Paetongtarn Shinawatra met Xi Jinping on February 6, 2025, and vowed to strengthen law-enforcement cooperation with China. By late February, Thailand allowed senior Chinese officials and Chinese security forces to participate in cross-border raids on scam centers in Burma.

Chinese government documents reveal that when Chinese security forces participate in raids on scam centers, they routinely confiscate large quantities of devices used in scamming operations. In August 2024, China's Ministry of Public Security announced that Chinese security forces had taken part in a raid in Burma and confiscated a large quantity of computers and phones, all of which were transported back to China.

These devices likely contain intelligence on Chinese criminal networks and highly sensitive personal data belonging to scam victims, including Americans. The U.S. Institute of Peace testified before the Commission that China has been unwilling to share information extracted from these devices with other countries.

On November 12, the U.S. Attorney's Office for the District of Columbia launched a new Scam Center Strike Force to fight the scam compounds operating across Southeast Asia. The operation brings together the U.S. Attorney's Office, DOJ's Criminal Division, the FBI, and the U.S. Secret Service, supported by the State Department, Treasury's Office of Foreign Assets Control, and the Department of Commerce. The Strike Force is pressing U.S. companies to cut these syndicates off from American digital infrastructure and shut down U.S.-based websites and accounts used to target victims.

Its Burma team has already moved against multiple compounds, seizing scam websites and seeking warrants to confiscate satellite terminals, while the FBI has deployed agents to Bangkok to work with the Royal Thai Police against notorious centers like KK Park.

Federal officials say the initiative will use every tool available to disrupt foreign criminal networks, recover stolen funds, and prevent U.S. systems from being weaponized against Americans. They hope the Strike

Force will also counter the CCP's influence and help stop Beijing from extracting even more data and money from American citizens.

CONCLUSION

China's deepening involvement in Burma since 2024 represents a comprehensive strategy to protect economic interests while keeping up plausible deniability about direct military involvement. Through technology transfers, surveillance systems, PSCs, and strategic mediation, Beijing has significantly strengthened the junta's capacity for repression while ensuring continued access to Myanmar's resources and strategic corridors.

This multi-faceted approach, combining military-technical cooperation, digital authoritarianism export, and economic leverage, demonstrates how China projects power and influence without deploying conventional military forces. As long as Beijing continues providing this support structure, the junta's ability to maintain power and suppress resistance will remain significantly enhanced, making prospects for genuine democratic transition increasingly distant.

METHODOLOGY AND RESEARCH STRATEGY

RESEARCH STRATEGY

This research adopted a qualitative approach, well-suited for analyzing China's complex and multifaceted involvement in Burma's ongoing civil war. Qualitative methods enable a detailed examination of China's political, economic, and military interventions, as well as the broader regional security implications of these actions.

A qualitative framework accommodates diverse paradigms such as post-positivism, interpretivism, and critical orientations, which allow for the exploration of contextualized and nuanced truths. As noted in *The Oxford Handbook of Qualitative Research* (Leavy, 2014), qualitative research often involves a reflexive engagement with sources, allowing researchers to construct partial truths that mirror the complexities of the subject. This approach is fitting for studying China's involvement in Burma, as this analysis engages with varied and sometimes conflicting sources, including government documents, media reports, and academic literature. By analyzing China's role from multiple perspectives — governmental, military, and humanitarian — this methodology enabled a nuanced understanding of how Beijing's interventions influence the conflict while accounting for potential biases in the sources. This approach captured the intricacies of geopolitical conflicts, like the Burma war, where contextualized interpretations are critical.

Through qualitative analysis, this study investigated China's support for the Burmese junta and various EAOs over time, demonstrating how these interventions have shaped the dynamics of the conflict.

DATA COLLECTION AND SECONDARY RESEARCH FOCUS

This study drew from a wide array of secondary data sources to analyze China's involvement in the Burma conflict. Sources included government documents such as official statements, treaties, and agreements from Chinese and Burmese officials, providing insight into China's role in the conflict. Reports from international organizations, including the United Nations and various aid groups, offered data on human rights abuses, displacement, and humanitarian crises connected to the conflict. Academic literature, including scholarly articles and books, was also examined to contextualize China's strategic objectives in Burma, including Belt and Road Initiative projects and regional security interests. Media coverage from regional and global outlets further complemented these sources by providing real-time analysis of China's ongoing interventions. All sources underwent qualitative content analysis to identify patterns and themes that revealed China's influence on Burma's economic, military, and diplomatic fronts.

SAMPLING CRITERIA AND CROSS-REFERENCING

To ensure accuracy and credibility, a cross-referencing approach was employed. By systematically verifying information across trusted sources, this process minimized inconsistencies and enhanced the reliability of the findings. Where possible, proxy data was also used for further comparison. For example, if the Burmese regime claimed to have exported goods worth $1 million to China, this figure was cross-referenced with China's import data for verification. Similarly, claims of Chinese investments or infrastructure projects in conflict zones were corroborated with sources from pro-junta media, EAO reports, and dissident media to provide a more comprehensive perspective.

Cross-referencing data on displacement, hunger, disease, and civilian deaths involved consulting reports from EAOs, cross-border aid organizations, and bodies like Amnesty International, the United Nations, and Human Rights Watch. This comparative approach allowed for a clearer understanding of the humanitarian impact of Chinese involvement and ensured that the research accurately reflected the on-the-ground situation.

DATA PROCESSING AND ANALYSIS PROCEDURES

The data processing involved a systematic coding method to identify and categorize recurring themes. Codes, represented by keywords or phrases, were assigned to segments of data to encapsulate central attributes or themes. These codes were organized into broader categories, allowing for comprehensive pattern analysis.

The main coding categories included:

- Economic Influence: Capturing China's investments in the China-Myanmar Economic Corridor, resource extraction projects, and financial support through loans to the Burmese junta.

- Military Support: Focusing on arms deals, provision of military equipment to the Tatmadaw, and support for certain EAOs.

- Diplomatic Engagement: Documenting China's role in ceasefire agreements, summits, and political negotiations that enhance the junta's legitimacy.

- Humanitarian Impact: Examining the adverse effects of Chinese-backed projects, including displacement and civilian harm.

Each code was applied by thoroughly reviewing source material, such as government documents and media reports, with tools including Excel used to organize and categorize data. Grouped codes were analyzed for patterns that indicate how China's economic, military, and diplomatic strategies shape the conflict.

CONTENT VALIDITY

Ensuring content validity involved operationalizing key constructs such as China's financial investments, military support, and diplomatic initiatives, verifying that the data collected from sources accurately represented these elements. For example, analyzing China's economic projects or military support to the junta offered concrete data aligned with these constructs. This approach strengthened the research's content validity by grounding findings in precisely defined variables (Marczyk, DeMatteo, and Festinger, 2005).

ETHICS

Ethical considerations were central to this research, especially given the sensitive nature of conflict data. The study exclusively relied on secondary sources, avoiding interviews, surveys, or firsthand accounts that could have placed participants in danger. By using publicly available documents and reports, the research ensured that individuals involved in the conflict faced no additional risk.

In handling sensitive data — particularly regarding human rights abuses and violence — care was taken to avoid unsupported conclusions. The potential biases in sources, especially those related to pro- or anti-junta positions, were acknowledged, and data was cross-referenced to enhance reliability. This approach respected the complexity of the conflict and maintained a commitment to objectivity, accuracy, and ethical research practices.

EXPECTED RESULTS FROM CODING

Through systematic coding, this research highlighted strategic patterns in China's involvement, including:

- Dual Support: Evidence that China supports both state and non-state actors to maintain influence on both sides of the conflict.

- Economic and Military Influence: Patterns showing that China's financial investments and military support help sustain the junta's power, thereby prolonging the conflict.

- Diplomatic Mediation: Insights into how China's diplomatic role legitimizes the junta while safeguarding its regional interests, such as energy security and trade route access.

FINDINGS
AND ANALYSIS

Overview of Chinese
Influence in the Burma Conflict

China's role in the Burma conflict is a manifestation of its broader strategic objectives, underpinned by economic ambitions and geopolitical needs. As shown in the literature review, China has maintained a multi-pronged approach in its engagement with Burma, positioning itself as a critical player both in supporting the Burmese junta and maintaining relationships with select EAOs. This influence is deeply tied to China's long-term goals, such as securing the CMEC for direct access to the Indian Ocean, which bypasses the Malacca Strait — key to addressing its "Malacca Dilemma." This infrastructure serves both economic and military purposes, enhancing China's regional leverage and securing alternative trade routes vital to its global supply chain.

Beijing's direct and indirect interventions have spanned financial support, diplomatic engagement, and military aid. Findings indicate that these measures work to stabilize the junta, undermine international sanctions, and protect Chinese investments in Burma. China's relationships with EAOs further exemplify its strategic balancing act. Supporting EAOs in northern Burma allows China to stabilize its border areas while countering groups that might otherwise oppose Chinese projects. Through this dual alignment with both state and non-state actors, China navigates the conflict in a way that maximizes its influence and ensures that any resolution or continuation of the conflict is compatible with its broader Indo-Pacific objectives.

Financial and Economic
Support for the Burmese Junta

China's financial and economic support for the Burmese junta plays a central role in sustaining the regime and advancing Beijing's economic interests. Findings indicate that Chinese investments in infrastructure projects such as the CMEC, oil and gas pipelines, and other development projects are critical to stabilizing the junta. These projects not only secure resource access for China but also deepen its economic ties to Burma, cementing the junta's dependence on Chinese funding and infrastructure.

Direct Investments and Infrastructure Projects

China's investment portfolio in Burma includes significant infrastructure projects that reinforce the junta's authority. The CMEC, for example, is a network of pipelines and transportation links that runs from the Burmese port of Kyaukphyu to China's Yunnan province. This corridor is strategically designed to give China direct access to the Indian Ocean, reducing its reliance on the vulnerable Malacca Strait. The financial support for these projects enables the junta to maintain its power, despite international sanctions that limit its other revenue streams. Findings reveal that this economic lifeline effectively undermines efforts by the international community to isolate the junta, as Chinese investments continue to flow.

Impact on Local Communities

While these projects benefit China and the Burmese junta, they have led to significant displacements and economic exploitation of local communities, particularly in ethnic minority areas. Evidence indicates that areas such as Rakhine and Shan states, which host major Chinese projects, have witnessed increased military presence and forced displacements. Local opposition to these projects is met with force by the Tatmadaw, often leading to human rights abuses. This correlation between Chinese-backed projects and intensified conflict highlights the disruptive effects of economic investments in regions where local populations are resistant to central government control.

Comparison to International Sanctions

Contrasting China's financial support with international sanctions reveals a clear counterbalance. While sanctions aim to pressure the junta by limiting its access to international capital, Chinese funding fills this gap. China's continued economic involvement undermines the effectiveness of sanctions, as the junta relies on Chinese investments to compensate for restricted access to Western markets. This relationship sustains the junta's economic viability and prolongs the conflict, as the sanctions fail to fully isolate Burma's leadership from external financial resources.

POLITICAL SUPPORT AND
DIPLOMATIC INVOLVEMENT

China's diplomatic support reinforces the junta's legitimacy while allowing Beijing to position itself as a stabilizing power in Burma. This political support takes the form of formal diplomatic ties, strategic mediation, and relationships with EAOs, which collectively serve to protect Chinese interests.

Beijing's political relationship with the junta is marked by regular meetings, public statements, and agreements that signal unwavering support. Findings indicate that Chinese officials frequently meet with Burmese leaders to affirm their partnership and mutual goals. This diplomatic engagement reinforces the junta's position, indirectly delegitimizing opposition forces and dissident movements. By positioning itself as a stable partner to the junta, China strengthens the regime's authority and counters international criticism aimed at Burma's military government.

China's alliances extend to certain EAOs, such as the UWSA and the MNDAA. The findings show that China has fostered relationships with these groups to maintain a balance of power in northern Burma, where its border stability and economic projects are concentrated. This dual support provides China with leverage over both the junta and EAOs, enabling it to act as an intermediary when necessary. By supporting both sides, China maximizes its influence and preserves the option to pivot depending on which faction offers the most stability or alignment with Chinese interests.

MILITARY SUPPORT AND STRATEGIC SECURITY GOALS

China's military support to Burma, though often indirect, has significant implications for the conflict's duration and intensity, as well as for regional security dynamics. Findings confirm that military aid from China enhances the junta's resilience, allowing it to counter both domestic resistance and international opposition. Beijing has reportedly provided the regime with various forms of military support, including arms sales and, at times, intelligence sharing. Instances of People's Liberation Army (PLA) troop sightings and Chinese weaponry in Burma reinforce the idea that China is willing to extend its support beyond economic and political means when necessary. This military assistance enables the junta to sustain operations against EAOs and maintain its grip on power, despite substantial resistance.

Border Security and Influence over Northern EAOs

China's support for EAOs near its border, particularly those in northern Shan State, serves the dual purpose of maintaining border stability and controlling potential security threats. The findings indicate that China backs these groups to manage instability at its border and to counter groups that may pose a risk to Chinese projects. This arrangement enables China to control the security dynamics in northern Burma, ensuring that its investments remain secure and that the conflict does not spill over into Chinese territory.

Implications for Regional Security

The implications of China's military support extend beyond Burma's borders, impacting US and Indian security interests in the Indo-Pacific region. China's growing influence in Burma, particularly through military means, reinforces its strategic posture in Southeast Asia. By securing alternative trade routes through Burma and preparing for potential US-China conflicts, China strengthens its regional dominance, potentially challenging India's influence and US interests in the area.

COMPLICITY IN THE ROHINGYA GENOCIDE

China's involvement in the Rohingya crisis has shown that its strategic interests take precedence over humanitarian concerns. Findings suggest that China's support for the junta, even during the Rohingya genocide, reflects its prioritization of stability and economic gains.

During the Rohingya genocide, China continued its diplomatic and economic support for the junta, even vetoing international resolutions against Burma. Beijing's political backing during this period allowed the regime to carry out mass displacements without significant international intervention. China's approach underscores a policy of non-interference in domestic issues, as long as they align with its strategic goals.

After the displacement of the Rohingya from Rakhine State, Chinese-backed projects moved forward with fewer local obstacles. Infrastructure initiatives, such as the Kyaukphyu Deep-Sea Port, benefited from the absence of displaced communities, who would have otherwise opposed the projects. This aligns with the argument that the removal of the Rohingya population facilitated China's investments in Rakhine, minimizing resistance to its strategic infrastructure projects.

IMPACT ON THE COURSE AND
DURATION OF THE BURMA CONFLICT

Chinese support has played a significant role in extending the Burma conflict. Findings show that Beijing's economic, military, and diplomatic backing enables the junta to endure external pressure and resist calls for democratic reform. By supplying the junta with resources, China mitigates the impact of international sanctions, prolonging the conflict.

China's involvement has also shaped the strategies of EAOs. Some EAOs have resisted Chinese-backed projects, while others have sought Chinese support, altering the power dynamics within the conflict. This divergence reflects how China's role influences both cooperation and conflict among EAOs, reshaping alliances based on proximity to Chinese interests.

Broader Strategic Implications for China's Regional Goals

China's investments in Burma serve its Belt and Road Initiative (BRI) by expanding its strategic footprint in Southeast Asia. Findings suggest that infrastructure projects like the CMEC enhance China's influence in the region, ensuring that Burma remains an integral part of its BRI strategy.

Burma's geographic position addresses China's "Malacca Dilemma," offering a direct route to the Indian Ocean and access to critical maritime pathways. Secure access to these routes positions China favorably in the event of US-China tensions, with Burma as a key link in its broader Indo-Pacific ambitions.

CONCLUSION

In conclusion, China's role in the Burma conflict intertwines economic, political, and military interests. The findings demonstrate that China's influence sustains the junta while simultaneously enabling its broader regional ambitions, particularly through its support of EAOs and also via projects like the CMEC. By providing financial investments, military aid, and diplomatic backing, China ensures the survival of the Tatmadaw despite international sanctions, prolonging the conflict and exacerbating the humanitarian crisis.

Economically, China's investments, particularly in infrastructure projects, serve its long-term strategic goals, such as addressing its "Malacca Dilemma" and expanding the Belt and Road Initiative. These initiatives not only reinforce the junta's authority but also displace local communities, intensify ethnic conflicts, and create conditions that suppress opposition to Chinese-backed projects. Politically, China's unwavering diplomatic engagement with the junta and selective support for EAOs illustrate a calculated strategy to maximize influence over both state and non-state actors, ensuring that any resolution to the conflict aligns with Chinese interests.

Militarily, China's provision of arms, intelligence, and drone technology bolsters the junta's ability to sustain its campaign against resistance forces, even as the Tatmadaw suffers territorial losses. Concurrently, China's support for EAOs near its border stabilizes its own security and protects key infrastructure projects. This dual alignment reveals China's prioritization of strategic stability and economic benefits over humanitarian concerns, as evidenced by its complicity in the Rohingya genocide and its facilitation of projects in regions affected by displacement.

China's involvement in the Burma conflict also reshapes the power dynamics within the region. By undermining international sanctions and empowering the junta, China delays democratic reform. Regionally, its activities in Burma solidify its influence in Southeast Asia and reinforce its Indo-Pacific ambitions, positioning Burma as a critical node in its geopolitical strategy. Ultimately, China's engagement in Burma secures its strategic objectives, but it comes at the cost of exacerbating conflict, displacing communities, and destabilizing the broader region. Without significant international intervention or a recalibration of Beijing's approach, the conflict in Burma is likely to persist, deepening the humanitarian crisis and further entrenching Chinese dominance in the region.

REFERENCES

Ahram, Ariel I., and Charles King. "The Warlord as Arbitrageur." *Theory and Society* 41, No. 2 (March 2012): 169-186.

Ahmad, Osama. "Ten Years of CPEC: A Decade of Disappointments." Stimson Center, August 18, 2023. https://www.stimson.org/2023/ten-years-of-cpec-a-decade-of-disappointments/.

Amnesty International. "Myanmar: Suspend Copper Mine Linked to Ongoing Human Rights Abuses." February 2017. https://www.amnesty.org/en/latest/press-release/2017/02/myanmar-suspend-copper-mine-linked-to-ongoing-human-rights-abuses/.

Asia News Monitor. "China/Myanmar (Burma): Narrow Chinese Interests Shaping Security Landscape in Northern Myanmar: US Expert." March 27, 2023.

Asia News Monitor. "Myanmar (Burma)/China: Myanmar Rebel Group Vows to Protect China's Interests." August 2, 2024.

Bertrand, Jacques, Alexandre Pelletier, and Ardeth Maung Thawnghmung. *Winning by Process: The State and Neutralization of Ethnic Minorities in Myanmar*. Cornell University Press, 2022.

Blue, Wayland. "Unrestricted Warfare Beyond the South China Sea: What We Can Learn about China's Pursuit of Strategic Influence from Its Engagement with Burma." *Marine Corps Gazette* 104, No. 11 (November 2020): 60-62.

BNI Online. "FPNCC: Best Friends, Worst Enemies." June 21, 2024. https://www.bnionline.net/en/news/fpncc-best-friends-worst-enemies.

Business & Human Rights Resource Centre. "Myanmar: Letpadaung Copper Mine Project Sparks Ongoing Protests, Land Disputes & Security Concerns." June 4, 2024. https://www.business-humanrights.org/en/latest-news/myanmar-letpadaung-copper-mine-project-sparks-ongoing-protests-land-disputes-security-concerns/.

Bynum, Elliott. "Analysis of the FPNCC/Northern Alliance and Myanmar Conflict Dynamics." ACLED, July 21, 2018. https://acleddata.com/archived/analysis-fpnccnorthern-alliance-and-myanmar-conflict-dynamics.

Cameron, Shaun. "Why Is Myanmar's New Deep-Sea Port Such Hot Property?" Lowy Institute, November 22, 2023. https://www.lowyinstitute.org/the-interpreter/why-myanmar-s-new-deep-sea-port-such-hot-property.

Chan, Debby. "How Important Is Lashio Militarily and Economically?" CNI Myanmar, October 4, 2024. https://cnimyanmar.com/index.php/english-edition/24923-how-important-is-lashio-militarily-and-economically.

Das, Bidhayak. "The Fall of Lashio, and the Road Ahead in Myanmar's Conflict." Border Lens, August 1, 2024. https://www.borderlens.com/2024/08/01/the-fall-of-lashio-and-the-road-ahead-in-myanmars-conflict/

DMG Newsroom. "FPNCC Welcomes Chinese Mediation in Myanmar's Affairs." DMG, March 16, 2023. https://www.dmediag.com/news/fpncc-welcomes-chinese.html.

Dukalskis, Alexander. "Why Do Some Insurgent Groups Agree to Cease-Fires While Others Do Not? A Within-Case Analysis of Burma/Myanmar, 1948-2011." *Studies in Conflict and Terrorism* 38, No. 10 (October 2015): 841-863.

Fan, Hongwei, and Yizheng Zou. "Rethinking the Security Issue in the China-Burma Territorial Dispute." *Cold War History* 23, No. 1 (January 2023): 45-59.

Findley, Michael G., and Josiah F. Marineau. "Lootable Resources and Third-Party Intervention into Civil Wars." *Conflict Management and Peace Science* 32, No. 5 (November 2015): 465-486.

Ganesan, N. "Myanmar-China Relations: Interlocking Interests but Independent Output." *Japanese Journal of Political Science* 12, No. 1 (April 2011): 95–111.

Jordt, Ingrid, Tharaphi Than and Sue Ye Lin. *How Generation Z Galvanized a Revolutionary Movement Against Myanmar's 2021 Military Coup*. ISEAS-Yusof Ishak Institute, 2021.

Kobayashi, Yuka, and Josephine King. "Myanmar's Strategy in the China–Myanmar Economic Corridor: A Failure in Hedging?" *International Affairs* 98, No. 3 (May 2022): 1013–32.

Kyi Sin. "Securing the China-Myanmar Economic Corridor: Navigating Conflicts and Public Scepticism – Analysis." *Eurasia Review*, October 7, 2024. https://www.eurasiareview.com/07102024-securing-the-china-myanmar-economic-corridor-navigating-conflicts-and-public-scepticism-analysis/.

Mandelkorn, Michael. *Arc of Complacency: The Rise and Fall of Soviet Monopoly over Burma*. Master's thesis, Tufts University. 2020.

Myanmar Now. "Myanmar junta troops torture locals after attack near Letpadaung copper mine." September 17, 2024. https://myanmar-now.org/en/news/myanmar-junta-troops-torture-locals-after-attack-near-letpadaung-copper-mine/.

Myers, Lucas. "The China-Myanmar Economic Corridor and China's Determination to See It Through." Wilson Center, May 26, 2020. https://www.wilsoncenter.org/blog-post/china-myanmar-economic-corridor-and-chinas-determination-see-it-through.

Myers, Lucas. "China's Economic Security Challenge: Difficulties Overcoming the Malacca Dilemma." *Georgetown Journal of International Affairs*, March 22, 2023.

Myitkyina Journal. "Junta – China Plan to Revive Myitsone Dam Project." BNI Online, June 10, 2024. https://www.bnionline.net/en/news/junta-china-plan-revive-myitsone-dam-project.

Naing, Ingyin. "Myanmar Conflict Unveils Complex Dynamics of China's Interests." Voice of America News, January 20, 2024. https://www.voanews.com/a/myanmar-conflict-unveils-complex-dynamics-of-china-s-interests-/7448050.html.

OEC. "China/Burma Trade." Accessed October 11, 2024. https://oec.world/en/profile/bilateral-country/chn/partner/mmr.

Ong, Andrew. *Stalemate: Autonomy and Insurgency along the China-Myanmar Border*. Cornell University Press, 2023.

RFA. "Chinese Workers Return to Copper Mines in Myanmar: Scores of Workers Arrive to Conduct Inspections of Equipment That's Been Dormant for Over 3 Years." Radio Free Asia, May 13, 2024. https://www.rfa.org/english/news/myanmar/chinese-workers-return-copper-mines-myanmar-05132024160209.html.

Rotberg, Robert I. (1998). *Burma: Prospects for a Democratic Future*. Brookings Institution Press, 1998.

Sahay, Navya. "CPEC: Is the Belt and Road Initiative's Crowning Project a Failure?" *Brown Political Review*, January 24, 2024. https://brownpoliticalreview.org/2024/01/china-pakistan-economic-corridor/.

Thein, Htwe Htwe and Sam Baron. "Business not as usual: International business and the Myanmar crisis." Asialink, University of Melbourne, February 15, 2024. https://asialink.unimelb.edu.au/diplomacy/article/business-not-as-usual-myanmar/.

Zhao, Hong. *China and India: The Quest for Energy Resources in the Twenty-First Century*. Routledge, 2012.

Zhao, Hong. *China and ASEAN: Energy Security, Cooperation and Competition*. ISEAS-Yusof Ishak Institute, 2015.

Zou, Yizheng, and Hongwei Fan. "The Political Economy of Welfare State and the China-Burma Relationship." *Cold War History* 19, No. 4 (October 2019): 569-585.

APPENDIX A:
COMPREHENSIVE LIST OF ARMED ORGANIZATIONS

A s of May 2025, Myanmar's civil conflict involves approximately 120-plus armed organizations ranging from major ethnic armed groups to local People's Defense Forces (PDFs). The conflict has dramatically escalated since the February 2021 military coup, with ethnic armed organizations (EAOs) now controlling significant territory across Myanmar's periphery.

I. MAJOR ETHNIC ARMED ORGANIZATIONS (EAOS)

A. Three Brotherhood Alliance (3BHA)

The most powerful coalition currently fighting the junta

1. Arakan Army / United League of Arakan (ULA)

- **Size**: ~45,000 troops (as of June 2024)
- **Location**: Rakhine State, expanding into Chin State, Magway, Bago, and Ayeyarwady regions
- **Alignment**: Independent (not formally aligned with NUG)
- **Status**: Active combat
- **Leadership**: Commander-in-Chief Major General Twan Mrat Naing, Vice Deputy Commander-in-Chief Brigadier General Nyo Twan Awng
- **Territory Controlled**: 13+ townships in Rakhine State, controls entire Myanmar-Bangladesh border (271km)
- **Notable**: Captured Western Regional Command headquarters in Ann (December 2024)

2. Myanmar National Democratic Alliance Army (MNDAA)

- **Size**: ~6,000 troops
- **Location**: Kokang region, northern Shan State
- **Alignment**: Independent (not aligned with NUG)

- **Status**: Ceasefire with junta (January 2025 - fragile)
- **Leadership**: Commander Peng Daxun (status uncertain; reportedly detained by China since late 2024)
- **Territory Controlled**: Kokang Self-Administered Zone, including Laukkai
- **Notable**: Captured Lashio (August 2024), withdrew as part of ceasefire deal (April 2025)

3. Ta'ang National Liberation Army (TNLA)

- **Size**: 10,000-15,000 troops (7 brigades, 30+ battalions)
- **Location**: Northern Shan State (governs Pa Laung Self-Administered Zone)
- **Alignment**: Cooperates with NUG-aligned forces, UNFC member
- **Status**: Active combat
- **Leadership**: Chairman Lt. Gen. Tar Aik Bong (PSLF), Commander-in-Chief Tar Hod Plarng, Secretary Tar Bone Kyaw
- **Territory Controlled**: Multiple townships in northern Shan State including Hsipaw, Kyaukme, Namhsan, Mogok, Pa Laung Self-Administered Zone
- **Notable**: Part of Operation 1027, captured Mogok (July 2024)

B. Northern Alliance Members

4. Kachin Independence Army / Kachin Independence Organization

- **Size**: ~20,000 troops (8 brigades)
- **Location**: Kachin State, northern Shan State, northwestern Sagaing Region
- **Alignment**: Cooperates with NUG and PDFs, FPNCC member

- **Status**: Active combat

- **Leadership**: Chairman General Gunhtang Gam Shawng (succeeded N›Ban La in January 2023), Commander-in-Chief Gen. Khaung Lun

- **Territory Controlled**: Eastern Kachin State, advancing toward Bhamo

- **Notable**: Captured 70+ military installations (March-April 2024)

C. United Wa State Army (UWSA)

Largest ethnic armed group in Myanmar

5. United Wa State Army (UWSA) / United Wa State Party (UWSP)

- **Size**: ~30,000 troops (Myanmar's largest non-state armed group)

- **Location**: Wa Self-Administered Division, Shan State

- **Alignment**: Officially neutral but supports some resistance groups

- **Status**: Maintains ceasefire with junta while supplying weapons to 3BHA

- **Leadership**: President/Commander-in-Chief Bao Youxiang (Tax Log Pang in Wa), Vice President Xiao Minliang, Deputy Commander-in-Chief Zhao Zhongdang & Bao Ai Chan (nephew of Bao Youxiang, promoted 2022)

- **Territory Controlled**: Wa State (de facto autonomous), expanding influence in northern Shan

- **Funding**: Business and trade with China, drug and weapon production

- **Notable**: Controls peacekeeping role in Lashio area, acquired surface-to-air missiles from China, does not seek independence

D. Eastern/Southern Groups

6. Karen National Liberation Army (KNLA) / Karen National Union (KNU)

- **Size**: ~15,000 troops (7 brigades)
- **Location**: Karen State (Kayin State), eastern Myanmar along Thai border
- **Alignment**: Cooperates with NUG and PDFs
- **Status**: Active combat
- **Leadership**: Chairman General Padoh Saw Kwe Htoo Win (elected May 2023), General Secretary Padoh Saw Tadoh Moo, KNLA Commander-in-Chief Gen Saw Johnny, Deputy Commander Lt-Gen Saw Baw Kyaw Heh
- **Territory Controlled**: Significant portions of Karen State (Kayin State), including Myawaddy area
- **Notable**: Myanmar's oldest EAO (fighting since 1949), signed NCA in 2015 but resumed fighting post-coup

7. Karen National Army (KNA) (formerly Karen Border Guard Force)

- **Size**: Unknown (several thousand)
- **Location**: Kayin State, primarily around Myawaddy
- **Alignment**: Initially junta-allied, then switched sides, now unclear
- **Status**: Complex relationship; has switched sides multiple times
- **Leadership**: Saw Chit Thu
- **Notable**: Split from Myanmar Army in January 2024, later facilitated junta troop movements

8. Karenni Army (KA) / Karenni National Progressive Party (KNPP)

- **Size**: ~2,000 troops

- **Location**: Karenni State (Kayah State)

- **Alignment**: Aligned with NUG through K3C alliance (see below)

- **Status**: Active combat

- **Leadership**: Chairman Khu Oo Reh, General Secretary U Aung San Myint

- **Territory Controlled**: Most of Karenni State (Kayah State) outside Loikaw

- **Notable**: Part of K3C alliance with KNU, KIA, CNF

9. Chin National Army (CNA) / Chin National Front (CNF)

- **Size**: ~8,000+ regular troops + ~10,000 auxiliary forces

- **Location**: Chin State, Kalay region

- **Alignment**: Aligned with NUG through K3C alliance, UNFC member

- **Status**: Active combat (ceasefire since 2012 but resumed fighting post-coup)

- **Leadership**: Chairman Pu Zing Cung, Vice-Chairman (1) Pu Thang Ning

- **Territory Controlled**: Large portions of Chin State

- **Notable**: Works closely with local PDFs and CDF groups

E. Mon and Southern Groups

10. Mon National Liberation Army (MNLA) / New Mon State Party (NMSP)

- **Size**: ~1,000+ troops

- **Location**: Mon State, parts of Karen State (Kayin State) and

Tanintharyi Region

- **Alignment**: Split - NMSP-Anti-Dictatorship faction (MNLA-AMD) opposes junta (formed 2024)

- **Status**: MNLA-AMD split from main group and declared war on junta (February 2024)

- **Notable**: Original NMSP signed NCA in 2015, but Anti-Military Dictatorship faction broke away

F. Shan State Groups

11. Shan State Army-South (SSA-S) / Restoration Council of Shan State (RCSS)

- **Size**: ~8,000 troops

- **Location**: Southern and eastern Shan State, Myanmar-Thailand border

- **Alignment**: Anti-SAC (resumed fighting post-coup despite NCA signatory status)

- **Status**: Active combat

- **Leadership**: Lieutenant General Yawd Serk

- **Headquarters**: Loi Tai Leng

- **Notable**: NCA signatory (2015) but resumed hostilities after 2021 coup, implemented mandatory conscription (6 years' service) for ages 18-45 in February 2024

12. Shan State Army-North (SSA-N) / Shan State Progress Party (SSPP)

- **Size**: ~10,000 troops

- **Location**: Northern Shan State

- **Alignment**: Anti-SAC (resumed fighting post-coup despite NCA signatory status)

- **Status**: Active combat, declared truce with SSA-S (November 30, 2023)
- **Notable**: NCA signatory but resumed combat after coup, planning to unite with SSA-S

13. Pa-O National Liberation Army (PNLA) / Pa-O National Liberation Organization (PNLO)

- **Size**: ~1,000 troops (active anti-junta faction)
- **Location**: Southern Shan State
- **Alignment**: Split - anti-junta faction aligned with resistance, pro-junta faction (PNLO-NCA-S)
- **Status**: Factional split in September 2024
- **Notable**: PNLO-NCA-S broke away to maintain ceasefire with junta

13b. Pa-O National Army (PNA)

- **Size**: ~4,000 troops
- **Location**: Southern Shan State
- **Alignment**: Fluid; some elements aligned with NUG, others with junta
- **Status**: Rearmed post-coup, ceasefire since 1991 but alignment uncertain
- **Notable**: Training 10,000 militia members to defend Naypyidaw approaches

G. Other Significant Groups

14. National Democratic Alliance Army (NDAA) - Mongla

- **Size**: ~3,000-4,000 troops
- **Location**: Eastern Shan State (Mongla area)

- **Alignment**: Part of FPNCC, maintains autonomy
- **Status**: Maintains ceasefire arrangements
- **Notable**: Controls Mongla Special Economic Zone

15. Lahu Democratic Union (LDU)

- **Size**: ~1,000-2,000 troops
- **Location**: Eastern Shan State
- **Alignment**: Generally neutral
- **Status**: Maintains low-level ceasefire

16. Lahu Democratic Union (LDU)

- **Size**: ~1,500 troops
- **Location**: Eastern Shan State
- **Alignment**: Anti-junta faction aligned with NUG (NCA signatory 2018)
- **Status**: Active against junta

17. Kayan New Land Party/Army (KNLP/A)

- **Size**: ~200-300 troops
- **Location**: Karenni State (Kayah State) State, southern Shan State
- **Alignment**: Pro-junta (allied with Tatmadaw since 1990s, maintains ceasefire)
- **Status**: Ceasefire with junta, fighting alongside military against resistance
- **Leadership**: Chairman Khun Than Tet
- **Notable**: Rival to KNPP, maintains ceasefire with military

H. New Post-Coup Armed Formations

18. Bamar People's Liberation Army (BPLA)

- **Size**: ~1,000+ fighters

- **Location**: Urban areas, central Myanmar

- **Alignment**: Aligned with NUG

- **Status**: Active resistance

- **Notable**: City-area resistance group formed post-coup

19. Mon State Revolutionary Force (MSRF)

- **Size**: Unknown

- **Location**: Mon State areas

- **Alignment**: Anti-junta

- **Status**: Active

II. PEOPLE'S DEFENSE FORCES (PDFs)

The PDFs represent the largest category of armed groups, with hundreds of local units across Myanmar.

A. National Unity Government (NUG) Organized PDFs

- **Total Estimated Size**: 100,000+ fighters across all regions (mid-2024 estimate)

- **Organization**: Five regional commands (Northern, Southern, Eastern, Western, Central)

- **Alignment**: Officially aligned with NUG

- **Funding**: Public donations, local taxation, NUG support

- **Structure**: Mix of regular PDFs (full military units), Local Defense Forces (LDFs - autonomous militias), and People's Defense Teams (PDTs - guerrilla groups)

- **Coverage**: Operating in 250 out of 330 townships nationwide

B. Regional PDF Commands

Northern Command

- **Location**: Sagaing Region, parts of Mandalay Region
- **Major Groups**: Sagaing PDFs, Mandalay PDFs
- **Estimated Size**: 15,000-20,000 fighters
- **Status**: Very active, coordinating with KIA and TNLA

Central Command

- **Location**: Central Myanmar (Magway, Bago regions)
- **Estimated Size**: 10,000-15,000 fighters
- **Status**: Active, working with various EAOs

Eastern Command

- **Location**: Shan State, Mon State
- **Estimated Size**: 8,000-12,000 fighters
- **Status**: Coordinating with ethnic groups

Western Command

- **Location**: Rakhine State, Chin State
- **Estimated Size**: 5,000-8,000 fighters
- **Status**: Working closely with AA and CNF

Southern Command

- **Location**: Karen State (Kayin State), Tanintharyi Region
- **Estimated Size**: 8,000-12,000 fighters
- **Status**: Integrated with KNLA operations

C. Notable Independent PDF Groups

Karenni Nationalities Defense Force (KNDF)

- **Size**: ~7,000+ troops (22 battalions, 6 brigades as of 2023)
- **Location**: Karenni State (Kayah State) State (Karenni State)
- **Alignment**: Works with KNPP/KA, loosely aligned with NUG
- **Status**: Very active
- **Leadership**: Chairman Khun Bedu (also vice-chair of Karenni State Interim Executive Council), Commander-in-Chief Major-General Aung Myat (also chief of KA)
- **Notable**: Strongest resistance group in Karenni State (Kayah State) State, formed May 31, 2021

Karenni National People's Liberation Front (KNPLF)

- **Size**: ~600 troops (BGF 1004 & 1005)
- **Location**: Eastern Karenni State (Kayah State) State (east of Salween River)
- **Alignment**: Defected from junta to resistance (June 2023)
- **Status**: Active against junta since June 2023
- **Leadership**: Spokesperson Lawren Soe
- **Notable**: Former Border Guard Force, officially joined resistance July 1, 2023

Kayan National Army (KNA)

- **Size**: 8 battalions
- **Location**: Western Karenni State (Kayah State) State, southern Shan State, parts of Karen State (Kayin State) and Naypyitaw
- **Alignment**: Recognizes NUG authority
- **Status**: Active (formed October 29, 2024)

- **Leadership**: Led by 14-person military committee of Karenni State Interim Executive Council
- **Notable**: Newest group, merger of Kayan ethnic forces

Chinland Defense Force (CDF)

- **Size**: ~5,000-8,000 fighters (multiple chapters)
- **Location**: Chin State, Sagaing Region
- **Alignment**: Works with CNF
- **Status**: Very active

Mandalay People's Defense Force

- **Size**: ~2,000-3,000 fighters
- **Location**: Mandalay Region
- **Alignment**: NUG-aligned, works with TNLA
- **Status**: Active in urban and rural operations

III. PRO-JUNTA FORCES

A. Myanmar Armed Forces (Tatmadaw)

- **Size**: ~150,000-200,000 (significantly reduced from pre-coup levels)
- **Status**: Suffering from defections, recruitment through conscription
- **Control**: Mainly urban centers and central Myanmar

B. Junta-Allied Militias

Shanni Nationalities Army (SNA)

- **Size**: ~1,000 troops
- **Location**: Kachin State
- **Status**: Allied with Tatmadaw and Shan State Army-South

Pa-O National Organisation (PNO) Militia

- **Size**: ~10,000 (being trained)
- **Location**: Southern Shan State (Hopong, Hsi Hseng, Pinlaung townships)
- **Status**: Training to defend Naypyidaw approaches, actively recruiting for Tatmadaw since 2021 coup
- **Leadership**: Armed wing of Pa-O National Organisation, allied with USDP

Karen National Army (KNA) - Note: Different from resistance KNA

- **Size**: ~7,000 troops
- **Location**: Kayin State
- **Status**: Formed in 2024 from border guard units, pro-junta
- **Notable**: Separate from the resistance Kayan National Army

Border Guard Forces

- **Size**: Various sizes
- **Location**: Multiple border regions
- **Status**: Subdivisions of Tatmadaw under Regional Military Commands
- **Notable**: Composed of former insurgent groups (Karen BGF, Kokang BGF, etc.)

Wuyang People's Militia

- **Size**: ~100 troops
- **Location**: Kachin State
- **Status**: Pro-junta

Pyu Saw Htee Militias

- **Size**: Unknown (village-level militias)
- **Location**: Various townships
- **Status**: Local defense forces

IV. ALLIANCE STRUCTURES

A. Federal Political Negotiation and Consultative Committee (FPNCC)

- **Led by**: UWSA
- **Members**: UWSA, NDAA-Mongla, SSPP, Three Brotherhood Alliance members (AA, MNDAA, TNLA), KIA
- **Status**: Non-NCA signatory alliance representing most pre-2021 non-ceasefire groups

B. K3C Alliance (Karen-Karenni-Kachin-Chin)

- **Members**: KNU, KNPP, KIA, CNF + NUG
- **Status**: Formal cooperation agreement (January 2024)

C. United Nationalities Federal Council (UNFC)

- **Members**: CNA, KNLA, and other groups seeking federal democracy
- **Status**: Umbrella organization for ethnic groups

D. Northern Alliance

- **Members**: KIA, TNLA, MNDAA, AA
- **Status**: Military cooperation

E. Three Brotherhood Alliance (3BHA)

- **Members**: AA, MNDAA, TNLA
- **Status**: Military alliance behind Operation 1027

F. Chin Brotherhood Alliance (CBA)

- **Location**: Chin State

- **Size**: 6 groups

- **Alignment**: Works with AA

V. TERRITORIAL CONTROL ASSESSMENT (MAY 2025)

Junta Control: ~30-40% of territory

- Central Myanmar (Yangon, Naypyidaw, Mandalay urban areas)

- Some state capitals (Loikaw in Karenni State (Kayah State), parts of other capitals)

- Major transport corridors (contested)

Opposition Control: ~35-45% of territory

- Most of Rakhine State (Arakan Army)

- Large portions of Chin State (CNF/CDF)

- Northern and eastern Kachin State (KIA)

- Much of Karen State (Kayin State) (KNU/PDFs)

- ~90% of Karenni State (Kayah State) (KNDF/KA/KNPLF/KNA coalition)

- Rural areas of Sagaing and Magway (PDFs)

Contested/Autonomous: ~15-25% of territory

- Wa State (UWSA - autonomous)

- Some Shan State areas

- Border regions

-

VI. KEY POLITICAL ORGANIZATIONS

A. Government-in-Exile

- **National Unity Government (NUG)**: Established 2021 by Min Ko Naing, includes ousted leaders Aung San Suu Kyi and Win Myint, recognized by EU, has representatives in USA and UK

- **Committee Representing Pyidaungsu Hluttaw (CRPH)**: 17 NLD lawmakers and parliamentarians, legislative body of NUG

B. Ruling Junta

- **State Administration Council (SAC)**: Led by Commander-in-Chief Min Aung Hlaing (also Prime Minister)

- **Union Solidarity and Development Party (USDP)**: Military-backed political party

- **Myanmar Armed Forces (Tatmadaw)**: Army, Navy, Air Force under Min Aung Hlaing

C. Dissolved Opposition

- **National League for Democracy (NLD)**: Led by Aung San Suu Kyi, won 2015 and 2020 elections, dissolved by junta

1. **Ethnic armies control most borderlands**

2. Junta losing territorial control rapidly

3. Increased cooperation between **EAOs and PDFs**

4. **Chinese pressure affecting northern groups**

5. **Arms production capabilities growing among resistance**

6. **Humanitarian crisis deepening with 3+ million displaced**

Note: Troop numbers are estimates based on available reporting and may vary significantly. The situation remains highly fluid with frequent changes in territorial control and alliance structures.

Sources: Based on research from multiple conflict monitoring organizations, news reports, and analyst assessments as of May 2025.

APPENDIX B:
CODING SPREADSHEET FOR QUALITATIVE DATA ANALYSIS

Coding Spreadsheet For Qualitative Data Analysis - Page 1

	Data Source	Excerpt	Code	Theme
1	Zhao (2015) - China and ASEAN: Energy Security, Cooperation and Competition	China's energy projects in Myanmar, including gas pipelines, secure	Energy Security Projects	Economic Influence
2	Blue (2020) - USMC Perspective on China's Influence in Burma	Chinese investments in infrastructure reduce dependency on Malacca Strait	Strategic Infrastructure	Economic Influence
3	Dukalskis (2011) - UWSA's Role in Chinese Border Security	UWSA operates autonomously along the China-Myanmar border, facilitating	Border Security Support	Military Support
4	Bertrand et al. (2022) - Neutralization of Ethnic Minorities in Myanmar	Ethnic armed organizations (EAOs) adapt to support or resist Chinese projects,	EAO Dynamics	Diplomatic Engagement
5	Amnesty International (2017) - Human Rights Concerns in Copper Mining Projects	Forced displacement for Chinese-backed mining projects impacts local	Forced Displacement	Humanitarian Impact
6	Asia News Monitor (2024) - MNDAA Pledges to Protect Chinese Interests	MNDAA commits to securing Chinese assets in Myanmar, highlighting	Local Alliance Protection	Military Support

China's Role in the Burma War: Jet Fuel, Money, and Weapons

Coding Spreadsheet For Qualitative Data Analysis - Page 2

	Data Source	Excerpt	Code	Theme
1	Ahmad (2023) - Stimson Center on CPEC Challenges	CPEC struggles highlight risks in China's foreign investments, relevant to Burma's	Investment Risks in Foreign Projects	Economic Influence
2	Findley and Marineau (2015) - Resource Conflicts	China's role in resource conflicts emphasizes its support for select actors to secure	Resource Conflict Involvement	Military Support
3	BNI Online (2024) - FPNCC Peace and Stability Efforts	The FPNCC, under Chinese guidance, engages in peace negotiations to stabilize border	Border Region Stabilization	Diplomatic Engagement
4	Cameron (2023) - Importance of Myanmar's New Deep-Sea Port	China's investment in Myanmar's ports aims to expand access to the Indian Ocean, reducing	Port and Trade Route Access	Economic Influence
5	Ganesan (2011) - China-Myanmar Political Relations	China maintains political ties with Myanmar's junta to counter Western influence and	Political Alliance	Diplomatic Engagement
6	Htwe and Baron (2024) - Ethical Dilemmas for International Companies in	Post-coup, international companies in Myanmar face ethical challenges	Corporate Ethical Challenges	Humanitarian Impact

Coding Spreadsheet For Qualitative Data Analysis - Page 3

	Data Source	Excerpt	Code	Theme
1	Cameron (2023) - Deep-Sea Port and Regional Control	China's investment in Myanmar's ports secures access to the Indian Ocean and bolsters	Port Access for Trade Control	Economic Influence
2	Findley and Marineau (2015) - China's Role in Resource Conflicts	China's interventions in Burma include securing resources through	Resource Conflict Partnerships	Military Support
3	Kobayashi and King (2022) - CMEC and Myanmar's Sovereignty	China-Myanmar Economic Corridor projects have raised concerns over Burma's	Economic Independence Concerns	Economic Influence
4	BNI Online (2024) - FPNCC's Role in Peace Negotiations	The FPNCC engages in peace negotiations with Chinese support, aiming to stabilize	China-Backed Peace Talks	Diplomatic Engagement
5	Myers (2023) - CMEC and China's Economic Security Needs	China's CMEC project highlights its push for alternative routes to secure economic	Alternative Trade Routes	Economic Influence
6	Amnesty International (2017) - Impact of Copper Mine Displacements	Local communities near copper mines face displacement and environmental degradation due to	Environmental and Humanitarian Impact	Humanitarian Impact

Coding Spreadsheet For Qualitative Data Analysis - Page 4

	Data Source	Excerpt	Code	Theme
1	Naing (2024) - Chinese Interests in Myanmar's Conflict	Chinese investments influence Myanmar's internal power struggles,	Influence on Power Dynamics	Economic Influence
2	Rotberg (1998) - Long-term Impact of China's Influence	China's engagement in Burma has historical roots, impacting the	Historical Political Impact	Political Influence
3	Htwe and Baron (2024) - Ethical Issues for Companies in Myanmar	Post-coup, companies are challenged with balancing profit motives and	Corporate Ethical Dilemmas	Humanitarian Impact
4	Ganesan (2011) - Political Support between China and Myanmar	China's political ties with Myanmar help reinforce the junta's power, countering	Political Support for Junta	Diplomatic Engagement
5	Sahay (2024) - Strategic Importance of CPEC and CMEC	CPEC and CMEC projects highlight China's strategy to circumvent traditional maritime	Alternative Trade Routes	Economic Influence
6	Jordt and Lin (2021) - Myanmar's Resistance Movements	Ethnic and resistance movements in Myanmar respond to Chinese	Local Resistance to Influence	Humanitarian Impact

Coding Spreadsheet For Qualitative Data Analysis - Page 5

	Data Source	Excerpt	Code	Theme
1	Das (2024) - Lashio Conflict and China's Role	China's involvement in Lashio underlines its interest in stabilizing conflict	Conflict Stabilization Efforts	Diplomatic Engagement
2	CNI News (2024) - Economic and Military Importance of Lashio	Lashio serves as a critical hub in China-Myanmar trade, where both military and	Strategic Trade and Military Hub	Economic Influence
3	DMG Newsroom (2023) - Chinese Mediation in EAO Peace Efforts	China actively mediates in EAO peace talks to secure its investments and	Mediation in Peace Efforts	Diplomatic Engagement
4	Fan and Yizheng (2023) - Historical Territorial Disputes	Historical border disputes have shaped China's ongoing strategies to manage	Historical Territorial Influence	Political Influence
5	Asia News Monitor (2023) - Narrow Security Interests in Northern Myanmar	China's narrow security goals focus on border stability, even if it means reinforcing the	Border Security Goals	Military Support
6	Kyi (2024) - Public Opposition to CMEC Projects	Public protests against CMEC highlight local concerns over environmental and	Local Opposition to Projects	Humanitarian Impact

Coding Spreadsheet For Qualitative Data Analysis - Page 6

	Data Source	Excerpt	Code	Theme
1	Ong (2023) - Opposition to Chinese CMEC Projects	EAOs oppose CMEC projects due to environmental impacts, disrupting China's investment	Environmental Opposition to CMEC	Humanitarian Impact
2	Cameron (2023) - Myanmar's Deep-Sea Port for Regional Access	China's port projects in Myanmar give it strategic control over access to the	Strategic Port Access	Economic Influence
3	Business & Human Rights Resource Centre (2024) - Letpadaung Copper Mine	Public protests over land confiscation near copper mines highlight the social cost of Chinese	Land Confiscation Protests	Humanitarian Impact
4	Zou and Fan (2019) - Burma as a Client State of China	Burma's reliance on China resembles a client state, dependent on Beijing's financial	Client-State Dynamics	Political Influence
5	Myanmar Now (2024) - Displacement Concerns around Copper Mining	Displacement of local populations due to mining activities raises humanitarian	Humanitarian Displacement	Humanitarian Impact
6	Bertrand et al. (2022) - Neutralization of Ethnic Minorities in	The Burmese government uses selective alliances with EAOs to	Alliance Management for Stability	Diplomatic Engagement

Coding Spreadsheet For Qualitative Data Analysis - Page 7

	Data Source	Excerpt	Code	Theme
1	Das (2024) - China's Role in Myanmar's Conflict Over Lashio	China's involvement in Lashio conflict highlights its focus on securing economic projects.	Economic Project Security	Economic Influence
2	Fan and Yizheng (2023) - Historic Border Disputes	Historic disputes shape China's modern policies, emphasizing control over border	Border Control Policies	Political Influence
3	Asia News Monitor (2023) - Security Concerns in Northern Myanmar	China's narrow security goals in Myanmar center around protecting strategic assets.	Strategic Asset Protection	Military Support
4	Blue (2020) - China's Strategic Influence Beyond South China Sea	China's interests in Burma align with its broader Indo-Pacific strategy for	Indo-Pacific Regional Influence	Political Influence
5	Sahay (2024) - CPEC Challenges and Strategic Interests	CPEC and CMEC reveal China's reliance on Myanmar for alternative trade	Alternative Trade Routes	Economic Influence
6	Kobayashi and King (2022) - Myanmar's Strategy with CMEC	CMEC projects bring Myanmar closer to Chinese economic influence, raising	Economic Sovereignty Concerns	Economic Influence

Coding Spreadsheet For Qualitative Data Analysis - Page 8

	Data Source	Excerpt	Code	Theme
1	Rotberg (1998) - Historical Context of Chinese Influence	China's historical influence in Myanmar affects the nation's current political alliances.	Historical Political Influence	Political Influence
2	Ganesan (2011) - China's Political Support for Myanmar Junta	China's political backing of Myanmar's junta strengthens the regime against …	Political Alliance Support	Diplomatic Engagement
3	BNI Online (2024) - FPNCC's Involvement in Peace Talks	The FPNCC, backed by China, is actively involved in peace efforts to stabilize the region.	Peace Stabilization Efforts	Diplomatic Engagement
4	Cameron (2023) - Control over Strategic Port Projects	China's investment in Myanmar's ports enables control over trade routes in the Indian Ocean.	Strategic Trade Route Control	Economic Influence
5	Htwe and Baron (2024) - Ethical Challenges for Companies in Myanmar	International companies face ethical dilemmas when operating near …	Corporate Ethical Dilemmas	Humanitarian Impact
6	Amnesty International (2017) - Land Confiscation for Mining	Forced land confiscations near mining projects bring humanitarian issues to light due …	Land Confiscation Issues	Humanitarian Impact

Coding Spreadsheet For Qualitative Data Analysis - Page 9

	Data Source	Excerpt	Code	Theme
1	Business & Human Rights Resource Centre (2024) - Letpadaung Copper Mine	Protests over the Letpadaung copper mine raise concerns about forced land confiscation by the	Forced Land Confiscation	Humanitarian Impact
2	Myers (2020) - China-Myanmar Economic Corridor (CMEC)	The CMEC provides a direct route to the Indian Ocean, aligning with China's broader	Direct Trade Routes	Economic Influence
3	Ahmad (2023) - CPEC and Investment Risks	CPEC's challenges highlight risks for Chinese investments, echoing potential	Investment Risk Reflection	Economic Influence
4	Kyi (2024) - Public Skepticism towards CMEC	Public skepticism and opposition toward CMEC projects reflect worries about	Public Opposition to CMEC	Humanitarian Impact
5	Sahay (2024) - CPEC as a Model for CMEC	CPEC's mixed success suggests that CMEC may face similar challenges in	Investment Model Comparison	Economic Influence
6	Myanmar Now (2024) - Displacement and Human Rights near	Displacement and human rights abuses near Chinese-backed	Human Rights Abuses	Humanitarian Impact

Coding Spreadsheet For Qualitative Data Analysis - Page 10

	Data Source	Excerpt	Code	Theme
1	Thant Aung (2024) - China's Ongoing Influence in Myanmar	China's influence in Myanmar remains strong as it continues to support projects	Sustained Regional Influence	Political Influence
2	Rotberg (1998) - Long-Term Sino-Burmese Relations	Historical ties between China and Burma underscore a legacy of political support for	Historical Strategic Support	Political Influence
3	Fan and Yizheng (2023) - Security Policies in Border Areas	Border security measures highlight China's aim to manage conflict spillover and	Border Security Measures	Military Support
4	Dukalskis (2011) - Ceasefire Agreements with EAOs	China has supported ceasefire agreements with EAOs to maintain stability along its	Ceasefire Support	Diplomatic Engagement
5	Jordt and Lin (2021) - Generation Z's Role in Resistance	Myanmar's Generation Z has mobilized resistance efforts that counteract	Youth-Led Resistance	Humanitarian Impact
6	CNI News (2024) - Strategic Importance of Lashio	Lashio is crucial for China due to its economic and logistical position	Economic and Logistical Hub	Economic Influence

GLOSSARY
AND ACRONYMS

GLOSSARY

Belt and Road Initiative (BRI): China's global development strategy aimed at enhancing regional connectivity and economic integration through infrastructure projects across Asia, Europe, and beyond.

China-Myanmar Economic Corridor (CMEC): A significant infrastructure initiative under the BRI, designed to connect Yunnan in China to Myanmar's Indian Ocean ports, enhancing trade and securing energy routes for China.

Ethnic Armed Organizations (EAOs): Armed groups in Burma/Myanmar formed by various ethnic minority communities. EAOs have historically fought for autonomy or independence and are involved in ongoing conflicts with the central government.

Malacca Dilemma: A term describing China's strategic vulnerability due to its dependence on the Strait of Malacca for energy imports. This reliance drives Beijing's interest in alternative trade routes, including the CMEC.

People's Liberation Army (PLA): The military arm of the Communist Party of China, responsible for national defense and advancing China's strategic interests.

Tatmadaw: The official name of the armed forces of Myanmar (Burma), which currently governs Myanmar as a military junta after the 2021 coup.

ACRONYMS

AA: Arakan Army, an ethnic armed organization in Myanmar.

AMU: American Military University.

BCP: Burma Communist Party, a historical political party and rebel organization that aimed to establish a communist state in Burma.

BRI: Belt and Road Initiative, China's international development and investment strategy.

CMEC: China-Myanmar Economic Corridor, a major BRI project connecting China to Myanmar's ports.

EAO: Ethnic Armed Organization, armed groups representing ethnic minorities in Myanmar.

FPNCC: Federal Political Negotiation and Consultative Committee, an alliance of ethnic armed organizations in Myanmar.

KIA: Kachin Independence Army, an ethnic armed organization in Kachin State, Myanmar.

KMT: Kuomintang, the Chinese Nationalist Party, which had forces in Myanmar after fleeing China following the Chinese Civil War.

MNDAA: Myanmar National Democratic Alliance Army, an ethnic armed organization based in northern Shan State, Myanmar.

NUG: National Unity Government, a parallel government formed by opponents of Myanmar's military junta.

PDF/PDFs: People's Defense Forces, anti-junta militias in Myanmar affiliated with the NUG.

PLA: People's Liberation Army, the armed forces of the People's Republic of China.

PRC: People's Republic of China.

SSPP: Shan State Progress Party, a political organization representing the Shan ethnic group in Myanmar.

UWSA: United Wa State Army, an ethnic armed organization in Myanmar with close ties to China.

INDEX